Management and Industrial Engineering

Series editor

J. Paulo Davim, Aveiro, Portugal

More information about this series at http://www.springer.com/series/11690

Carolina Machado
Editor

Competencies and (Global) Talent Management

Springer

Editor
Carolina Machado
School of Economics and Management
University of Minho
Braga
Portugal

ISSN 2365-0532 ISSN 2365-0540 (electronic)
Management and Industrial Engineering
ISBN 978-3-319-53398-8 ISBN 978-3-319-53400-8 (eBook)
DOI 10.1007/978-3-319-53400-8

Library of Congress Control Number: 2017930650

Printed on acid-free paper

This Springer imprint is published by Springer Nature
The registered company is Springer International Publishing AG
The registered company address is: Gewerbestrasse 11, 6330 Cham, Switzerland

Preface

Nowadays, it is common to hear that organizations need to be highly committed in order to hire, manage, and retain talented employees able to answer to the competitive challenges that characterize current markets. Talent management, or if we are focusing an international and global context, global talent management, is one of the key issues of the business strategy, with which all organizations are concerned. As a business strategy issue, (global) talent management must be completely integrated within all of the employee-related processes of the organization. At this level, managers and/or the human resources managers have a relevant role and responsibility in the recruitment process as well as in the ongoing development of and retention of superior employees. They are responsible to find, retain, and develop a superior workforce, with high levels of competencies and talent. While in some organizations only top potential employees are included in the talent management system, in others everybody is included in this process. The ideal is that everybody in the organization could be developed in order to obtain high levels of competencies becoming a talent. The existence of effective HRM and HR information systems, with adequate information flows, is very important to the performance and success of the organization as they contribute to track the career paths of the employees as well as to manage the available opportunities for competent and talented employees.

Taking into account this reality, this book, entitled **Competencies and (Global) Talent Management** looks to cover the issues related to the competencies and talent management in the modern and competitive organizations, in six chapters, namely:

First chapter presents "Global Talent Management in the Not-for-Profit Sector"; second chapter contains information about "The Transferable Skills Development Programme of a Portuguese Economics and Management Faculty: The perceptions of Graduate Students"; third chapter covers "Organizational Propensities to Share: Revisiting Talent Mobilization and Redistribution in Multinational Corporations"; fourth chapter discusses "What Is Talent Management? The Perception from International Human Resources Management Students"; fifth chapter focuses on "Job Market, Generations and Talents"; finally, in sixth chapter, "Global Talent

Management: Reality or Utopia? A Special Glance Through a Portuguese Multinational Organization" is presented.

Giving a special attention to the development of competencies and talent management, this book is a relevant contribution to help managers become better prepared to face the challenges and changes that today's organizations are facing with their workforce. In other words, it looks to provide a support to academics, researchers, engineers, as well as those that operating in the management field need to deal with policies and strategies related to the workforce issues.

The editor acknowledges her gratitude to Springer for this opportunity and for their professional support. Finally, I would like to thank to all chapter authors for their interest and availability to work on this project.

Braga, Portugal Carolina Machado

Contents

Editor and Contributors

About the Editor

Carolina Machado received her Ph.D. degree in management sciences (organizational and politics management area/human resources management) from the University of Minho in 1999 and master's degree in management (strategic human resource management) from Technical University of Lisbon in 1994. Teaching in the human resources management subjects since 1989 at University of Minho, she is since 2004 associated professor, with experience and research interest areas in the field of human resource management, international human resource management, human resource management in SMEs, training and development, management change and knowledge management. She is head of Human Resources Management Work Group at University of Minho as well as chief editor of the International Journal of Applied Management Sciences and Engineering (IJAMSE).

Contributors

Sandra Araújo EDIT VALUE Formação Empresarial, Braga, Portugal

Iris Barbosa University of Minho, School of Economics and Management, Braga, Portugal

Chris Brewster Henley Business School, University of Reading, Reading, UK

Jean-Luc Cerdin ESSEC Business School, Cergy, France

Carla Freire University of Minho, School of Economics and Management, Braga, Portugal

Delfina Gomes School of Economics and Management, University of Minho, Braga, Portugal

Carolina Machado School of Economics and Management, University of Minho, Braga, Portugal

Sara Oliveira EDIT VALUE Formação Empresarial, Braga, Portugal

Joana Ribeiro School of Economics and Management, University of Minho, Braga, Portugal

João Leite Ribeiro School of Economics and Management, University of Minho, Braga, Portugal

Mariana Paiva Santos University of Minho, School of Economics and Management, Braga, Portugal

Kushal Sharma ESSEC Business School, Cergy, France

David Starr-Glass University of New York in Prague, Prague, Czech Republic; Empire State College, International Programs (Prague), State University of New York, Prague, Czech Republic

Global Talent Management in the Not-for-Profit Sector

Chris Brewster, Jean-Luc Cerdin and Kushal Sharma

Abstract This chapter explores the issue of competencies and global talent management in an usual setting: not-for-profit organisations. Detailed examples from intergovernmental organisations (members of the United Nations family) and a non-governmental organisation (*Médecins Sans Frontières*) provide evidence for us to argue that these organisations have both special advantages and special problems in talent management. Their advantages include the levels of commitment to the mission and the focus on beneficiaries; their problems include the non-competitive salaries and the unwillingness to spend resources that could go to the beneficiaries on the staff. We outline the distinction between these organisations and the more commonly studied for-profit organisations and suggest learning that could take place between them.

1 Introduction

This chapter examines talent management (TM) in not-for-profit organisations (NFPs), with a focus on mission-driven organisations (see [1]). To begin, we first need to define what we mean by talent management and explain the characteristics of mission-driven organisations.

Talent management has become a widely used term among human resource management (HRM) practitioners, consulting firms and professional associations. There is, however, no consensus about the exact definition of the term [2]. Broadly speaking, there are 'exclusive' and 'inclusive' approaches [3]. The exclusive, or elitist, definition proposes that TM is a set of policies and practices aimed at

C. Brewster (✉)
Henley Business School, University of Reading, Reading, UK
e-mail: c.j.brewster@henley.ac.uk

J.-L. Cerdin · K. Sharma
ESSEC Business School, Cergy, France

C. Machado (ed.), *Competencies and (Global) Talent Management*,
Management and Industrial Engineering, DOI 10.1007/978-3-319-53400-8_1

managing a relatively small group of employees whom the organisation considers as essential for its superior performance. It is argued that the 'talents' are only a small proportion of the workforce who exert a significant impact on the performance of the organisation—these are usually key technical specialists, senior management and other individuals with high potential [4]. The inclusive approach, on the other hand, assumes that all, or at least most, of the employees have talent and that the task is to manage them all in ways that will bring out and develop those talents: this brings the definition very close to that of previous definitions of HRM. Regardless of the approach, TM focuses on the attraction, identification, development, retention and deployment of talent [2, 5, 6]. It looks to the future as well as being concerned about the present, aiming to groom potential future leaders [7]. The notion that organisations need to invest extra time and energy to retain employees who 'rank at the top in terms of capability and performance' [8] is not new but has become increasingly popular since organisations increasingly believe that individuals who are disproportionately more productive than their co-workers [9] can help them to gain competitive advantage [10]. Multinational enterprises (MNEs) in particular have realised that they have to 'spread their net' beyond locally available talent, thus connecting TM with the additional complexity of global HRM [11].

By far, the majority of TM work has focused on those businesses whose *raison d'être* is profit. Mission-driven organisations are different. NFPs do operate within the capitalist system and though they are not beholden to shareholders, they must operate efficiently enough to make sure that their outgoings do not exceed their incoming financial resources. But for NFPs, money is just a means to an end. They hope to be able to use their money for other purposes, and critical questions are raised if they have too much surplus of income over expenditure: they have what has been called a 'distribution constraint' [12]. They are judged, by themselves and others, on how far they are able to make progress towards their mission. The missions can be very varied. Some of these organisations have huge and highly laudable missions (the United Nations [UN] aims for 'world peace'; the European Union [EU] aims to bring the states of Europe together into an ever closer union; the Save the Children Fund aims to 'save children's lives. We fight for their rights. We help them fulfil their potential'). Other NFPs have more modest missions ('to make sure no one in our area is alone at Christmas'). Some NFPs are part of the public sector, charged with carrying out the administration of the public systems and with meeting certain objectives aimed to ameliorate or improve citizens' health, welfare or well-being. Others are deliberately non-governmental, they may be supported by government grant aid, by wealthy individuals or the public in general, and their objectives may range from encouraging musical appreciation to the preservation of a language, to running local sports competitions. Some are controversial: organisations trying to preserve the privileges of particular ethnic groups, or to advance particular religions, or to change drug legislation, for example.

TM in NFPs will differ from TM in profit-driven organisations. In the relatively new TM field, this is one of the many areas of inquiry that have been insufficiently explored. In this chapter, we address this gap and analyse TM in mission-driven organisations. Our aim is to identify the main features of TM in NFPs with the intention of opening up new avenues for future research. We restrict ourselves to TM in NFPs that organise across national borders. Such organisations might recruit someone in one country, train them in another, allocate them to work in a duty station in another country, and then promote or reassign them to yet another geographic location. TM with an international component has been labelled global talent management (GTM), which has been defined as a set of activities falling within International human resource management concerned with aligning HRM with the organisation's strategic direction and attracting, developing, retaining and mobilising talent [13]. The definition is taken from profit-focused MNEs but fits the context of NFPs too and realistically captures their TM activities. Managing talent globally includes expatriation [14] and other forms of global mobility management [15]. Through our case studies, we explore how far NFPs make use of talent spread across different geographical locations.

2 Diversity of Talent Management in Mission-Driven Organisations

This section uses three case studies to examine the diversity of TM in NFPs. There will be differences in the TM policies of NFPs depending on the nature of the work they do. Our case studies are drawn from two main sources of data: (a) information provided by organisational members through semi-structured interviews and (b) information posted on the organisations' respective websites. We cover the following:

- TM at *Médecins Sans Frontières* (MSF)
- TM at the United Nations Children's Fund (UNICEF)
- Joint TM at United Nation's Office for the Coordination of Humanitarian Affairs (OCHA) and the United Nations Development Program (UNDP).

We use the same structure for each case:

1. We begin with a brief overview of the organisation;
2. We consider how the organisation defines talent and talent management;
3. We discuss the organisation's approach to attracting, developing and retaining talent;
4. We explore the uses of technology to facilitate the TM process;
5. We present our interviewees' views about the future of TM at their organisation and the challenges that they need to overcome to establish TM as the formal system for managing talented employees.

3 TM at MSF

3.1 Organisation Overview

Médecins Sans Frontières (MSF) is comprised of 24 associations, coordinated by MSF International, whose secretariat is based in Switzerland. It is a neutral non-governmental body, largely funded by private donations. Established in 1971 with the purpose of delivering emergency medical care to individuals caught in crisis due to natural disasters, armed conflict and epidemics, MSF does not take sides in armed conflicts, but does speak out publicly where it finds that aid systems are not being utilised adequately or are being diverted for political reasons. It employs a large number of health professionals and logistical and administrative staff to achieve its mission. MSF employs both expatriates and predominantly locally hired staff.

3.2 Definition of Talent and TM

TM has just been introduced at MSF and is in an early phase of implementation without, as yet, formal rules, practices, and policies. According to one of our interviewees:

> …people have integrated what they believe the organisation does…it is something very implicit and sometimes very informal.

This informal approach of identifying talents leads to a lack of diversity, evidenced by the fact that most expatriates at MSF are Europeans. A more decentralised system, more based on explicit criteria and with greater staff ownership, would help to overcome that.

MSF's HRM policies have traditionally been focused on ensuring that it has the people and the skills necessary to carry out its work in the field and that it does not fall foul of local employment laws. Most MSF staff (90% or more) are contracted locally, so recruitment is decentralised. Staff are usually employed for one project and they leave at the end of the project. Therefore, turnover is high, with MSF continuously starting new projects. Recruitment is, therefore, an ongoing activity. Since it is hard to predict the length of many projects, employees usually get one-year renewable contracts.

Recently, MSF has realised that it needs to invest in talent and TM to reduce problems such as high turnover:

> …development of our people is important because everyone else is doing it…So let's just say that it started out as…as just comparisons with other organisations…and right now, it is starting to matter but they are also starting to understand what it is and how it can be developed in MSF.

Many employees are unaware of how the organisation manages talent, since TM is not formalised, so that even people identifying talent might not be aware that what they are doing is a part of TM. And talents are often subjectively evaluated. Supervisors might recommend someone for a mission, but the criteria for such recommendations are implicit:

> …the trouble is that…nobody confronts anyone else on what the criteria are…they say 'oh look, I think this person will be highly suitable to go and take up this job' [but] based on what criteria? What is it that we are looking for?

According to another interviewee, the organisation perceives talents as those who have gone on more than one or two missions. She adds that retention is a serious problem, which means investing a lot of resources into a person who only goes for one mission does not make much sense.

At MSF, TM is exclusive and informal but some in the organisation want it to be inclusive.

3.3 Main Focus of TM

To assure its employees of fairness, MSF wants an objective system outlining and communicating the desired skills and comparing employees' existing skills with them:

> Actually, what we are aiming for is to have something a little bit more structured where we all agree on some criteria…Otherwise, it becomes too subjective. We need to identify the ones that are particularly promising, that we want to invest in…

TM initiatives at MSF were started originally with the aim of improving succession planning:

> We are … an emergency organisation … so it is hard to convince anyone to do anything if it is not urgent… succession planning…is very important and very quickly becomes urgent if someone leaves. So this is how we framed it. But, in the future, my vision of TM would be something that is much more … But we are still very far from that.

One of the interviewees, who had worked for other not-for-profit organisations, argued that practice is becoming closer to that of the private sector, in areas such as talent identification and succession planning. She argued that TM at MSF is still based on defining competence on the basis of experience, whereas TM at the NFP where she worked before was more structured and explicitly communicated. In her opinion, big corporations have even more structured programmes, with better resources and desired employee behaviours outlined:

> MSF does not want to put all its people in a box. It is quite happy to have the people who do not always necessarily conform …so they have some difficulty in saying…we need people who display behaviour A, B and C…They say [that] we want people who express [themselves] and if this is their personality…then so be it. So they are less normative.

A particular problem of TM at MSF is to manage dual-career couples. In the medical profession, particularly, it is common to find both partners involved in medicine but with each engaged in their own careers. The challenge with expatriate assignments is to accommodate the career needs of the partner. For MSF, it is rare that both partners are doctors and/or nurses, which would make it easier to move them together. Furthermore, MSF wants to ensure that its employees are committed to their work rather than being dependent on their partners. It has been argued that trailing partners play an important role in ensuring the success of expatriate managers [16] and although there may be reasons to be suspicious of these suggestions [17] they do have an influence. However, since MSF assignments might require expatriates to go to conflict-ridden areas, it is not always possible for expatriates to take their family along on their assignments.

3.4 Attracting Talents

MSF realises that salary and benefits are usually major factors that attract candidates to work for an organisation. However, matching compensation and benefits to external markets is not a priority for MSF. They pay well below the going market rate, so that the organisation is clear that people are coming for the work, not the cash:

> We can't really afford to be competitive [with the private sector] but we do at least want to make sure that we give people conditions that are sufficiently compatible … [so that] they are willing to stay… people come also for the value of the work and that in a sense there has to be a balance.

But although MSF is not looking to compete with other organisations on the basis of salary, it still needs to attract talent. At MSF, according to one of our interviewees, two reasons for the attractiveness of the organisation are the opportunity of receiving a high-responsibility post even at a very young age, and the fact that employees can have a say in where they want to be deployed.

3.5 Developing Talents

MSF identifies talents generally informally but also through formal performance appraisals and annual seminars. HR managers offer those who have been identified as talents new opportunities in the form of bigger or more complex jobs, including:

> senior managers and/or the people we would consider might be senior managers in 1-3 years…we look at how they are doing [in] a particular job. So, for example, would they be ready to take a senior…job? So we look at the skills required…and then we [identify] who seems to be most promising.

Most of the training at MSF is on-the-job training including technical as well as managerial training. In addition, there are seminars:

> ...we will have one week with them where ... they can express their concerns or their opinions and their ideas...but they also learn from each other and they learn from us...and we learn from them...this is very important to their professional development.

3.6 Retaining Talents

Retention is the biggest challenge for MSF. The 'selfless rationale' works for many people at first, but it is difficult to maintain for an extended period, particularly when employees start a family. Retention applies to both local talent and expatriates. International mobility is often associated with increased stress and less support. MSF clearly states on its website:

> Living conditions in the field (security, housing, etc.) make it difficult for field staff to be accompanied by children or partners who are not participating in the program.[1]

Thus, expatriates might need to leave their partner and children behind and this becomes increasingly difficult over time. According to one of our interviewees:

> Once they have children...they really think hard whether they want to continue. I was just talking to a colleague here this morning and I said, 'Would you go back to the field?' and he said, 'listen, now I have a little boy...I don't want that kind of a life'. So we lose a lot of people for this reason.

Since other organisations offer better compensation and less stressful and risky work environments, there is always somewhere for the talent to go. Medical staff who have worked at MSF are highly marketable: recruiters are impressed by people who have, at least for a time, sacrificed salary and working conditions to help people in risky and sometimes dangerous situations. MSF's website clearly states the following:

> To reach those who need our help the most, we often work in conflict and post-conflict regions. MSF staff may live and work in insecure environments.[2]

MSF is aware of the problems involved in retaining talents. Our interviewees suggest this can be improved through better performance management and more dialogue with people regarding their performance, in order to avoid people quitting simply because no one from HRM has taken the time to sit down with them for a conversation. Other interviewees believe that retention can be improved by acknowledging the efforts made by its employees:

---

[1] http://www.msf.org/work-msf/working-in-the-field. Accessed 07 May 2014.

[2] http://www.msf.org/work-msf/before-you-apply. Accessed 07 May 2014.

> This is not an environment where there is a lot of recognition, you know…you do a good job and people think 'yeah, that's great' but they won't necessarily say so. They are more prompt to see what doesn't work. So I think we need to reverse that and…give much more recognition even to how hard people are trying because it's a very exhausting type of work…

3.7 Use of Technology

Our interviewees felt that some of MSF's systems are outdated and noted that performance records are kept on excel sheets, so they are hard to analyse.

In training and development, MSF does provide some e-learning opportunities, but our interviewees were worried about the impersonal nature of e-learning:

> [we] have some e-learning platforms…but, there is always a problem of e-learning in terms of, you know…they are a little bit closed and…impersonal - you are just working in front of a machine…even if they put a system of tutors and so on…to make it a little more interactive… it is not the ideal way of working.

Some of these problems could not be addressed because of financial constraints:

> The budget probably is very minimal if existent at all, in terms of what can be done around talent management in the organisation.

3.8 Future of TM

Several factors strengthen the TM-oriented approach at MSF. The continual flow of external candidates brings new perspectives. While many individuals who have been with MSF for a long time believe that turnover cannot be avoided, these new employees see this as a problem to be solved. Even existing employees, when they study for higher degrees in HRM-related fields, become aware of TM practices at other organisations and come up with alternatives that might work better than the existing arrangements. One of our interviewees thought TM would be systematic in future both in headquarters and in the different countries where MSF works, although she thought that MSF needed to change its attitude to TM:

> So far the organisation has considered that this work is HRM…it's HRM's job to find the people and put them in positions. And so, managers have not invested much time in considering in a very formal way who are the talents…and really, take some kind of ownership in ensuring that people grow…We are trying to overcome [this].

MSF provides long-term contracts to very few people, usually only for administrative jobs in its headquarters. One of our interviewees expected that TM would allow MSF to manage turnover of talent better in the near future. She thought that

MSF needs to give long-term contracts to a selected number of people so that they can envision a future with MSF. This applies especially to people who have been sent out as expatriates, who deserve a chance for their work to be recognised, by being offered a long-term contract.

3.9 *Challenges in Implementing TM*

With the current subjective criteria system, managers do not have to justify their decision to recommend someone as a talent. A minority of managers prefer it to remain that way, as it enables them to recommend people they like. In MSF, key stakeholders will need to be convinced that TM produces beneficial outcomes for the organisation. As with most HRM initiatives, the immediate benefits of TM are not apparent:

> It's like planting a seed. You have to wait…you have to put water…and you have to protect it…but it takes time…And people here are very impatient because they are working in emergencies and, you know, you don't save lives with talent management!

4 TM at UNICEF

1. ***Organisation overview***

   Established in 1946, UNICEF is a UN agency with a broad mandate for children. It advocates for children's rights—the right to survival, the right to education, the right to gender equality, and the right to protection. It works closely in partnership with governmental and non-governmental organisations, civil society and community-based organisations.

2. ***Definition of talent and TM***

   UNICEF has what they call their New and Emerging Talent Initiative (NETI), instigated in 2008. NETI is an entry point for young professionals looking for a career with UNICEF. Initially, the idea was to encourage young people to join UNICEF at junior and mid-level positions for two years, with individuals showing good performance being offered regular staff positions with a fixed-term contract. Since then, as one of our interviewees said:

> The initiative has evolved a lot…today it includes both … national staff who are interested in moving into international career plus external candidates who just want to join the organisation through the programme.

UNICEF's official website outlines the conditions of eligibility for the NETI programme, listing minimum educational qualifications, desired competencies and preferences regarding the age of the candidates. It does not provide an explicit

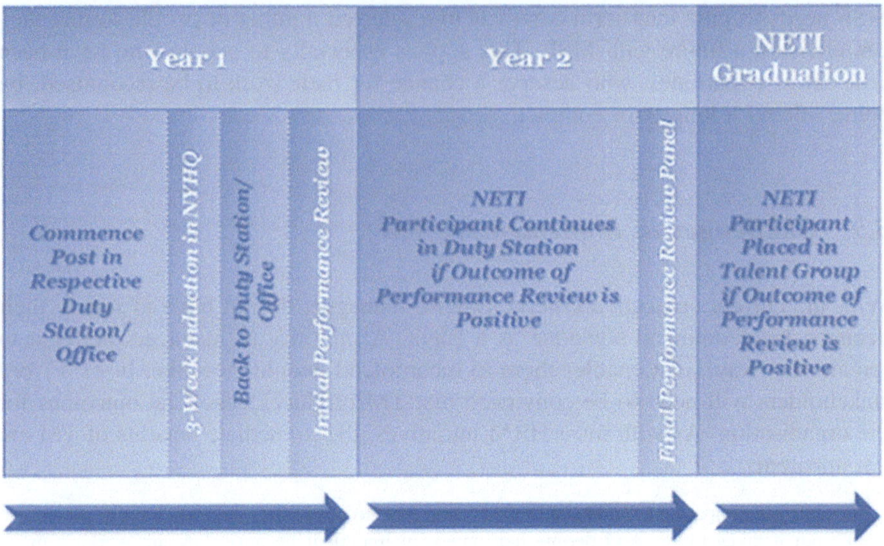

Fig. 1 Progression of NETI participants. *Source* http://www.unicef.org/about/employ/index_74417.html

definition of talent.[3] Participants receive a three-week orientation at UNICEF's headquarters in New York[4] and then start work in a duty station for a year. If they receive a positive evaluation at the end of the year, their contract is extended for a second year. About 18 months from the date of their first appointment, NETIs undergo another review process to decide whether they should be included in the Talent Group and by implication, appointed to a permanent contract. Figure 1 is taken from the UNICEF website.

In UNICEF, TM is part of the organisational unit 'Staff Mobility and Rotation' within the human resources division, so that NETI is integrated with the mobility of international professional staff. NETI is one, albeit important, strategy within the overall staff recruitment, mobility and rotation process.

In UNICEF, 'talent', according to one of our interviewees, is whoever meets the requirements of the job, survives the first two years, and is assessed as a good performer:

> If I look at the way the talent pool itself is built up…a talent is whoever fits the requirement of that specific area of work in terms of generic job description…[You] then need to prove it over a period of time…and if you are able to survive the first two years and you can manage in the third one, then you have been good.

[3]http://www.unicef.org/about/employ/index_74422.html.

[4]http://www.unicef.org/about/employ/index_74417.html.

3. *The organisation's approach to attracting, developing and retaining talent*

Many employees in the UN system are long-serving. New employees are more likely to import new ideas and new attitudes towards work. Interviewees stressed that investment in talent management is essential to sustaining the organisation.

TM is also focused on succession planning. The thinking is:

> how do we make our succession planning as smooth as possible? [We] need to be able to fill the gap as [people] move on and, frankly speaking, these new blood being injected in the organisation, they are really, really welcome and…we have to adjust.

For UNICEF, an important target of TM is to ensure a pool of candidates who can be deployed whenever there is a need. This pre-vetted applicant pool is called the 'Talent Group'—applicants who have gone through a vetting (or selection) process and meet UNICEF's criteria. Applicants from this pool are given priority for vacancies as they have already been screened, helping UNICEF to cut down on the time spent on hiring candidates to meet emergency needs. For example:

> During the Ebola crisis, we needed to deploy people to work in affected countries…and I would say that thanks to the talent pool we were able to respond swiftly and better than many other organisations because we could list a number of candidates who are suitable and we could offer them the job and they could be deployed in 48 h.

Interviewees believed that compensation and benefits are important for attracting candidates in any organisation. UN staff earn more than those in most other NFPs, but their salary and rewards are less than that of their equivalents in MNCs. Comparing the attractiveness of UNICEF's reward system with that of MNCs, one of our interviewees said:

> [In MNCs] it's all around bonuses… it's money-driven. If you do something outstanding, you expect the organisation to pay you a bonus or to give you some extra pay… We don't have bonuses. You do whatever you do, you do it well…you do an outstanding job…yes, there is recognition, but it's not in monetary terms.

However, UNICEF has little trouble in attracting candidates. Many people find the humanitarian mission of UNICEF, with its concern for children, very appealing:

> We have people who dream and really would love to join the organisation because of its mandate for children. I would say that it doesn't really cost UNICEF much to attract attention and to…you know…create interest for people to join.

Generally, UNICEF does not need to take special measures to attract talent. There was a campaign a few years back, where the human resources director talked to students in schools about working at UNICEF, but this is not something UNICEF does regularly.

UNICEF utilises a combination of on-the-job as well as formal, virtual, and off-the-job training methods, so that employees have a range of options in terms of development, such as training:

> with different educational institutions, including Harvard. We have quite well developed learning and development policies which allow staff members to be part of a continuous learning programme…We also allow staff the flexibility to leave the organisation temporarily on special leave with or without pay to go and take the programme they want for their own development.

The challenge for UNICEF is not attraction but retention. UNICEF makes a substantial investment in hiring and developing people, so retaining them makes economic sense for the organisation. However, many leave. Some find they do not fit with the organisation's culture and some find better career opportunities.

4. ***Use of technology***

UNICEF's centralised e-recruitment system allows the organisation to build a pool of candidates for potential employment opportunities and retains applicants' personal, educational, and work-related data for two years.

They do not, however, have a system to track the progress of their talents. If they could:

> reach a level where we can properly monitor how these NETI participants are performing, how are they being mentored and coached and so on…then what we get at the end of the two years…that would definitely help a lot.

4.1 Future of TM

Our interviewees expressed a belief that performance management should be a part of TM, because they feel that the performance management system is not streamlined enough to be able to really monitor talent. In the current system, some people may be managed as talents because of their initial screening, even though their subsequent performance indicates that the screening was inaccurate. It was also felt that TM could improve the performance management system by, for example, helping to set clear performance expectations:

> Unfortunately … the tool itself is not the best tool on earth. It has a lot of shortfalls and it makes it difficult for a NETI to feel comfortable after his/her performance evaluation because the supervisors don't necessarily attach a lot of importance to the tool. They don't apply it properly.

4.2 Challenge of TM

The main challenge for TM at UNICEF is to overcome the already established norms and culture. People in the organisation were reluctant to discuss performance problems because, as one interviewee expressed, they are as tight-knit as a family.

4.3 Joint TM at OCHA and UNDP

The unique feature of the TM programmes of two United Nations (UN) organisations, the United Nations Office of the Coordination of Humanitarian Affairs (OCHA) and the United Nations Development program (UNDP), is that it is a joint programme. Both organisations need to identify and develop the best leaders for more than 130 key coordination positions in their field operations. OCHA has been working for a number of years on an interagency model of TM. The participants of the TM programme are not OCHA's staff members, and thus OCHA does not have full control over them:

> ...how do you do Talent Management in an inter-agency manner when the talent is owned by the individual agencies ... you cannot promise them anything for the future because that all happens at the agency level.

4.4 Organisation Overview

OCHA was created in 1991 with the mandate of ensuring efficient responses during humanitarian crises through the coordination of all stakeholders involved in relief operations. When a crisis arises, the Chief of OCHA or Emergency Relief Coordinator (ERC) appoints a humanitarian coordinator (HC). The HC ensures that the needs of the affected population are addressed without the duplication of efforts or waste of resources. UNDP employs resident coordinator (RC)/resident representative (RR) roles in most developing countries. Sometimes, in a smaller country, the same person can hold more than one of these roles.

4.5 Definition of Talent and TM

To identify people for those jobs, competencies are defined and an assessment is designed and conducted by an independent entity. Twice a year, organisations within the UN system can nominate their senior leaders to go through a rigorous, independent and transparent Resident Coordinator Assessment Centre (RCAC). Depending on their availability to take up a position outside their parent organisation, successful RCAC candidates are put onto a roster. These candidates apply for upcoming coordinator positions and are then further screened by a UN Inter-Agency Appointment Panel (IAAP) based on their assessment reports, their updated CVs, and in some specific cases further interviews. From among the shortlisted candidates, the UN Secretary General selects the final candidates for RC/RR/HC positions.

OCHA, on behalf of the Inter-Agency Standing Committee (IASC), has supported the development of a specific talent management programme, called the IASC HC Pool, where candidates from the UN and non-UN agencies (NGOs or Governments) are pre-screened and rostered by the IASC panel members based on humanitarian competencies.

4.6 Main Focus of TM

The joint programme started in 2009 with the aim of building a pool of candidates for the role of HC and selecting the right candidate when the need arose. One of our interviewees pointed out that the screening and interview and being placed in the 'potentials' pool are just the first step:

> that doesn't guarantee that they are going to be deployed… they can apply for positions. It is a roster … it is not for a job in particular … We do this HC pool process once a year or once every two years, and we replenish the pool…

The RC/RR of the UN is usually tasked with the additional role of humanitarian coordinator. However, RC/RRs are typically experienced in coordinating long-term processes rather than responding to an immediate and urgent need, so there is a need to identify RC/RRs who are versed not only in socio-economic development but also in humanitarian coordination issues. The OCHA office, as Secretary to the IASC, has worked with UNDP to attract, identify, and train humanitarian leaders to mainstream in the RC/RR career track. OCHA also works with other organisations to select the best candidates. According to one of our interviewees:

> The humanitarian committee is composed of not only UN agencies but also of NGOs and the Red Cross movement… it is our duty to involve all the parts of the humanitarian community. So basically what we are trying to do is to identify the best and the brightest to go and become the humanitarian coordinators - that is the top official coordinating humanitarian responses, leading and coordinating humanitarian responses in the field.

TM programmes normally examine their own organisation's workforceto identify best performers, but the TM programmes in this case strengthen the coordination mechanisms among a variety of organisations with different mandates and interests, but the same broad goals. Candidates inside and outside the UN are examined to find those with the best competencies to lead coordination efforts with credibility among an increasing number of stakeholders in socio-economic development and humanitarianism. So, although they look at different pools of candidates as well as at different consensus-building stakeholders, both TM programmes aim at identifying similar competencies. By working together on talent management, UNDP and OCHA are increasing the effectiveness of the overall leadership coordination in both the socio-economic development and the humanitarian field.

4.7 Attracting Talents

The IASC HC Pool process provides a free and safe assessment against the HC competencies which are similar to those of the RC/RR. This allows the UN to test their own candidates prior to sending them to an expensive assessment centre such as the RCAC. Not many agencies are using the IASC HC Pool as a pre-requirement for the RCAC, though. For non-UN candidates, the IASC HC Pool process not only acknowledges their humanitarian coordination leadership but most importantly it also builds consensus around candidates who are not known to the UN. Thus, both UN and non-UN organisations have incentives to nominate their candidates to this pool.

For UN candidates, UNDP RC/RR/HC positions are senior positions—at a level where equivalent roles within the UN system are rare—and thus these posts are often the final step in their careers. They may for a few, of course, be key stepping stones to a very high political appointment elsewhere. Candidates find these positions attractive because of their high rank and status.

However, attracting candidates might be more of an issue in the future. These RC/RR/HC positions may be losing their appeal, since they are only temporary in nature and administrative hurdles make it difficult to understand what may happen once the assignment has been completed:

> When you go into one of these positions, you usually get promoted… however, you remain a member of your organisation of origin…they will take you back [but] at your previous grade…so it's actually a demotion…that limits the attractiveness of this career track.

4.8 Developing Talents

OCHA also provides mentoring, coaching, specific training and performance advice to the RC/RRs appointed in countries where a crisis may arise. Successful candidates are promoted to more complex and prestigious positions, usually above their current level of duties, and get exclusive training, access to privileged information and access to important networks for their future career. However, the promotion only lasts for the duration of their assignment, and the role serves the entire UN community beyond their organisation of origin.

OCHA has also designed development programmes especially for senior women managers. The UN is better than almost any comparable organisation in ensuring gender equality, applying positive discrimination and insisting on shortlists that contain at least one woman. While significant progress was made among the RC/RRs, the percentage of women among HCs remains low. It is thought that this is due at least in part to the hardship conditions of these positions, including unsafe locations and 'not-for-family' duty stations. OCHA, on behalf of the IASC, is paying management consultancy Deloitte to pilot a programme targeting senior women managers with experience in socio-economic development issues for temporary assignments in humanitarian postings, to allow them to acquire the necessary experience. At the same time, senior women managers with humanitarian experience are sent to socio-economic development agencies to broaden their competencies. Selected staff continue to be paid by their own organisation and return there after one year. Although a long-term investment, the programme aims at encouraging more women to apply for these humanitarian leadership roles. ODHA argues that even if they do not take this career path, the enlarged networks acquired by both sides will be beneficial for the overall coordination of UN field operations and possibly a catalyst for more partnerships and cooperation between the socio-economic development and the humanitarian world.

4.9 Retaining Talents

Retaining talent is not an issue for the joint TM programme, which is intended to be temporary in nature.

4.10 Use of Technology

Equally, the limited number of candidates suitable to be included in the HC pool means that managing them does not require the use of sophisticated technology.

4.11 Future of TM

These two TM programmes for the whole UN system offer screening, interviewing, shortlisting and recommendations for the RC/RR/HCs as the result of a consensus-building process involving the stakeholders. Hence, building and sustaining the credibility of the individuals selected is not an imposed decision but a broad consensus activity for which all organisations can take credit. One issue worrying OCHA is the potential inconsistency in interview panels since the same panel members do not interview all candidates.

One of the chief benefits of these programmes is that, in the highly political environment of the UN system, they provide a structured and unbiased process for identifying leaders. The biggest problem is that the programmes are not linked to TM programmes within the participating UN or non-UN organisations. The rationale is that careers are seen as an individual responsibility, unrelated to the needs of any particular organisation within the humanitarian family. According to one of our interviewees, such problems are more numerous for non-UN organisations:

> ...someone else has taken your place basically...all your peers have moved up, ... and when you try to come back, you are not considered part of the family anymore...so this is a huge problem.

In the UN and its agencies, nearly all positions are advertised and internal candidates have to undergo written tests and interviews even if they have already served in a similar position or have been identified as suitable for posts at an upper level. Nominations of prospective candidates to the two described TM programmes in UNDP and OCHA are, as a result, often driven by recruitment deadlines rather than by careful organisational TM decision-making. Sometimes, also, participating organisations do not want to lose their top talent:

> ...the agencies want to retain their top talents...[so] do not necessarily nominate the best person...or anybody at all. And the fact that...agencies have to nominate...people cannot apply on their own, is a major bottleneck.

4.12 Challenges of TM

This joint TM programme, with several stakeholders and reliant on coordination between several different organisations, has many complexities and challenges, of which perhaps the biggest is creating acceptance for non-UN candidates. For them, the IASC HC Pool programme is one of the few entry points to leadership positions in the UN system. However, as 'unknowns' they have problems competing with existing UN employees. UN organisations compete to get their candidates selected and actively lobby for their candidates in positions relevant to their mandates, making it more difficult for those without such lobbying behind them. However, the success of the few non-UN candidates who made it to these positions encourages other external candidates to try this career path.

There is also a challenge to justifying to stakeholders why non-UN organisations should benefit from these UN-provided, and paid for, programmes. The increasing presence of new actors in the relief arena and increased acceptance of non-UN candidates is important for the smooth coordination of humanitarian operations, but increases the pressure to justify this expenditure, especially given the ongoing financial constraints on the UN. Non-UN organisations may not have the financial means to cover the additional cost of training, travel, and other related expenditures for their successful candidates. As non-UN candidates need to be sponsored by a UN organisation (e.g., OCHA), in a time of financial constraints (see Chapter "What is Talent Management? The perception from International Human Resources Management Students"), it may be difficult to justify to donors why a UN agency is sponsoring a non-UN candidate who may not stay within the system.

Another challenge is to maintain the neutrality of the system. OCHA is mandated to organise these activities, but the interagency nature of the IASC HC Pool requires a 'firewall' to ensure the neutrality of serving on the IASC. This may be problematic as OCHA is also a key decision-making member of both Inter-agency TM processes described.

5 Main Differences Between Talent Management in NFPs and MNCs

There is very little literature on TM in NFPs, so to draw out the differences between them and the more frequently studied MNCs, we derive the characteristics of TM in the NFPs from our case studies and compare that with the existing academic literature on TM in MNCs.

6 Comparison of TM in MNCs and NFPs

For such comparisons, we follow the same structure in which we presented our case studies: how organisations define talent and talent management; their approach to attracting, developing and retaining talents; their use of existing technology to facilitate TM; and finally we discuss the future and challenges for TM.

6.1 Definition of Talent and TM

As noted, organisations can take either an inclusive or an exclusive approach to managing talent: Do a 'gifted' few possess talent or does everyone possess developable strengths [18: 1779]? Thus, there is a controversial side to managing talents. There are few, if any, objective criteria for defining who is a talent and who is not. Indeed, identification as 'talent' may be a self-fulfilling prophesy. This subjectivity raises ethical and moral questions, especially in organisations that promote egalitarianism [19]. Managers may also need to take into account the prevailing culture of the organisation and the views of employees. As a result, some organisations engage in TM practices but do not label them as such [20: 1841]. We found evidence of both inclusive and exclusive approaches to TM in our cases: UN organisations took an exclusive approach, whereas MSF wanted an inclusive approach.

6.2 Main Focus of TM

Organisations can have different reasons for engaging in TM. In MSF, TM started as a way of improving succession planning. This is not uncommon among organisations—succession planning is near the top of the agenda for both NFPs and

MNCs [7]. However, even when NFPs were aware of the steps they needed to take to manage talent better, often cost constraints acted as a barrier, including the fact that NFPs are considerably more limited in the reward packages they can offer employees. MNCs have less constraints and can make large investments to achieve the best results.

6.3 Attracting Talents

The 'war for talent' has attracted considerable attention from academics and practitioners, suggesting that there is a shortage of talent so that organisations must compete for the best employees [10, 21]. Arguably, this is self-serving language that deliberately ignores millions of people who do not fit a very narrow model of talent, but both NFPs and MNCs believe that they may be in such competition. Looking from the other side of the fence, individuals want meaningful careers and may be less prepared to do so within their current organisation [22]. 'Employer branding' [23] can be a powerful tool for NFPs to convey their HRM policies in terms of career managementand ultimately of talent management.

6.4 Developing Talents

Developing the skills of their most valued employees is good for the individuals and also at least in the long run for the organisation. In NFPs, most training and development directed towards the talents was on-the-job training. However, the link between good performance and opportunities to engage in training and development is not straightforward, since the criteria for judging performance are complex; more so than in MNCs where the 'bottom line' is always a basic fallback. For both kinds of organisations, of course, there are subjectivity issues. As McDonnell and [24: 59] noted, social capital—'the ability to build and sustain relationships and networks'—determines who is labelled a talent and therefore is provided with developmental opportunities.

6.5 Retaining Talents

Retention is one of the major challenges of TM. Preece et al. [19] found that a combination of factors including high turnover created talent shortages in Chinese subsidiaries of some MNCs. While NFPs rarely find it difficult to attract talents, because of their missions, retaining them in the long run can be more of a problem —particularly in the NGO sector. Even when talented people are initially satisfied, they may later look for other opportunities and options. Many sectors of the private sector experience similar problems. Money is not always the answer: NFPs attract and retain talent even though their salary and benefits are low compared to MNCs.

6.6 Technology

NFPs rarely have the latest e-enabled support systems. Hence, they are not able to exploit employee and performance data. We noted, for example, that UNICEF cannot track the progress of NETI participants and MSF uses Excel sheets for performance appraisals. MNCs tend to have well-developed systems to track employee performance as well as the career progression of talents.

6.7 Future of TM

MNCs have apparently been ahead of NFPs in engaging in the 'war for talent' [21]. However, we expect to see an increase in the popularity and adoption of TM in the NFPs: they clearly have less experience than MNCs, and some of our interviewees felt they could help reduce this gap by learning about TM philosophy, tools and practice from the MNCs.

Since both types of organisations operate in many countries, they both seem to accept the need for a global component to existing TM. MNCs have an advantage here, given the cost constraints in NFPs. For example, MSF persists with local hiring partly because it reduces costs, even though they realise the advantages of developing talent through expatriation and other forms of international mobility. This is less of an issue for MNCs' GTM practices. MNCs are utilising and experimenting with different forms and sources of talent such as self-initiated expatriates (see [25]), skilled migrants [26, 27], inpatriates [28, 29] and short-term assignees [30].

6.8 Challenges of TM

These are early days of TM in NFPs. One of the key challenges is to convince stakeholders of the long-term benefits. Another is to overcome resistance from powerful organisational members who have grown used to existing ways of working. A third challenge is to establish objective criteria for identifying, developing, and rewarding talent. Unlike MNCs, NFPs have to face a greater degree of complexity, sometimes political complexity. One challenge for MNCs is to manage the jealousy and lack of cooperation from those not included in the talent pool [31], but this may be less of an issue in some of the governmental and intergovernmental organisations where the identification of talent depends on the results of rigorous examinations open to all.

7 Mutual Learning Between NFPs and MNCs

Although issues such as retaining talent may be common to both NFPs and MNCs, their approach to managing such issues differs. An important reason lies in the divergence between them in terms of vision, mission and goals.

Nevertheless, NFPs and MNCs can learn from each other in order to manage their talent better. In this final section, we make recommendations for both types of organisations.

- *Attracting talents*: NFPs need to learn from MNCs that focusing on attracting the right people is crucial, but due to limited resources in comparison with MNCs, they need to use different weapons in the 'war for talent'. Their advantages lie in their meaningful missions and the opportunities that provide for their employees to have careers that 'make a difference'. There is room here for MNCs to learn from NFPs that the mission and values of an organisation may also be vital attractors.
- *Developing talents*: We recommend NFPs provide more developmental opportunities. This will involve increased investment in the staff, always difficult for organisations with moral missions to spend their money on. Career management and development programmes are beneficial to both types of organisations, but are often lacking in both.
- *Retaining talents*: Retaining talent is seen as a problem in both NFPs and MNCs. Turnover is a larger problem for short projects. In our opinion, MNCs can learn from NFPs that the contribution one makes to society is as important to employees as the compensation and benefits. Thus, we would recommend MNCs to attempt to create meaningful jobs with a sense of purpose to retain their talents.
- *Use of technology*: NFPs can learn from MNCs in relation to exploiting available employee information. MNCs tend to be good at maintaining and using information on their employees, while NFPs often struggle with rather outdated sources and systems with data often not being readily available. Since having information about the performance of talents is crucial to TM, we recommend NFPs make use of better tools for managing and updating information about their talents.
- *Future of TM*: TM at NFPs is in the early stages of development, compared to MNCs. NFPs can learn from the experience of MNCs and should take steps to do so. They may need to catch up too with the ongoing discussions in the field of TM and especially in GTM—the idea that an organisation needs to make use

of all available sources of talent across the world and not rely only on local talent. By contrast, MNCs can learn from the more 'internationalist' mindset common in NFPs.

- *Challenges of TM*: Important stakeholders of NFPs are recognising the need for TM. Benchmarking their own programmes against successful TM programmes in MNCs would be valuable, and their benefits could be used to promote and gain support for TM in the NFPs.

8 Conclusion

Because of the scarcity of academic literature on TM in NFPs, we had to conduct case studies in order to understand how NFPs manage TM. Even from this limited evidence, it is clear that there is extensive diversity of TM in NFPs, as there is in MNCs. From the extant TM literature and the analysis of our case studies, we identified the main differences between TM approaches of NFPs and MNCs. Our research is very far from exhaustive, but it does provide a useful indicator for a largely un-researched area—TM in NFPs. Thus, this chapter not only contributes to the TM field by expanding our knowledge to less studied organisations, but may also help organisations to improve practice. We hope that future research will build on the foundation laid by this chapter to further explore and expand our knowledge of talent management in not-for-profit organisations.

References

1. Cerdin, J.-L., Sharma, K., Tarique, E., & Purpura, C. (forthcoming). Talent management in mission driven organisations. In C. Brewster & J.-L. Cerdin (Eds.), *HRM in mission driven organisations*.
2. Lewis, R. E., & Heckman, R. J. (2006). Talent management: A critical review. *Human Resource Management Review, 16*(2), 139–154.
3. Stahl, G., Björkman, I., Farndale, E., Morris, S., Paauwe, J., & Stiles, P. (2012). Six principles of effective global talent management. *MIT Sloan Management Review, 53*(2):25–32.
4. Festing, M., Budhwar, P. S., Cascio, W., Dowling, P. J., & Scullion, H. (2013). Current issues in International HRM: Alternative forms of assignments, careers and talent management in a global context. *German Journal of Human Resource Management, 27*(3), 161–166.
5. Collings, D. G., & Mellahi, K. (2009). Strategic talent management: A review and research agenda. *Human Resource Management Review, 19*(4), 304–313.
6. Tansley, C., & Tietze, S. (2013). Rites of passage through talent management progression stages: an identity work perspective. *International Journal of Human Resource Management, 24*(9), 1799–1815.
7. Mäkelä, K., Björkman, I., & Ehrnrooth, M. (2010). How do MNCs establish their talent pools? Influences on individuals' likelihood of being labeled as talent. *Journal of World Business, 45*(2), 134–142.

8. Stahl, G.K., Björkman, I., Farndale, E., Morris, S. S., Paauwe, J., Stiles, P., et al. (2007). Global talent management: How leading multinationals build and sustain their talent pipeline. INSEAD Faculty and Research Working Papers, 2007/24/OB.
9. Groysberg, B., Lee, L.-E., & Nanda, A. (2008). Can they take it with them? The portability of star knowledge workers' performance. *Management Science, 54*(7), 1213–1230.
10. Mellahi, K., & Collings, D. G. (2010). The barriers to effective global talent management: The example of corporate élites in MNEs. *Journal of World Business, 45*(2), 143–149.
11. Sparrow, P., Brewster, C., & Chung, C. (2016). *Globalizing Human Resource Management* (2nd ed.). London: Routledge.
12. Hansmann, H. (1980). The role of non-profit enterprise.*Yale Law Journal, 89*, 835–901.
13. Tarique, I., & Schuler, R. S. (2010). Global talent management: Literature review, integrative framework, and suggestions for further research. *Journal of World Business, 45*(2), 122–133.
14. Cerdin, J.-L., & Brewster, C. (2014). Talent management and expatriation: Bridging two streams of research and practice. *Journal of World Business, 49*(2), 245–252.
15. Collings, D. G. (2014). Integrating global mobility and global talent management: Exploring the challenges and strategic opportunities. *Journal of World Business, 49*(2), 253–261.
16. Harvey, M. (1998). Dual-career couples during international relocation: The trailing spouse. *International Journal of Human Resource Management, 9*(2), 309–331.
17. Haslberger, A., & Brewster, C. (2008). The expatriate family—An international perspective. *Journal of Managerial Psychology, 23*(3), 324–346.
18. Sparrow, P., Farndale, E., & Scullion, H. (2013). An empirical study of the role of the corporate HR function in global talent management in professional and financial service firms in the global financial crisis. *International Journal of Human Resource Management, 24*(9), 1777–1798.
19. Preece, D., Iles, P., & Jones, R. (2013). MNE regional head offices and their affiliates: Talent management practices and challenges in the Asia Pacific. *International Journal of Human Resource Management* (April 2013), 1–21.
20. Valverde, M., Scullion, H., & Ryan, G. (2013). Talent management in Spanish medium-sized organisations. *The International Journal of Human Resource Management, 24*(9), 1832–1852.
21. Minbaeva, D., & Collings, D. G. (2013). Seven myths of global talent management. *The International Journal of Human Resource Management, 24*(9), 1762–1776.
22. Arthur, M. B., & Rousseau, D. M. (Eds.). (1996). *The boundaryless career: A new employment principle for a new organisational era*. Oxford: Oxford University Press.
23. Martin, G., & Cerdin, J. L. (2014). Employer branding and career theory: New directions for research. In P. Sparrow, H. Scullion, & I. Tarique (Eds.), *Strategic talent management: Contemporary issues in international context* (pp. 151–176). Cambridge: Cambridge University Press.
24. McDonnell, A., & Collings, D. G. (2011). The identification and evaluation of talent in MNEs. In H. Collings & D. G. Scullion (Eds.), *Global talent management*. New York, London: Routledge.
25. Jokinen, T., Brewster, C., & Suutari, V. (2008). Career capital during international work experiences: Contrasting self-initiated expatriate experiences and assigned expatriation. *International Journal of Human Resource Management, 19*(6), 979–998.
26. Al Ariss, A., Koall, I., Özbilgin, M., & Suutari, V. (2012). Careers of skilled migrants: Towards a theoretical and methodological expansion. *Journal of Management Development, 31*(2), 92–101.
27. Cerdin, J.-L., Abdeljalil Diné, M., & Brewster, C. (2014). Qualified immigrants' success: Exploring the motivation to migrate and to integrate. *Journal of International Business Studies, 45*(2), 151–168.
28. Cerdin, J. L., & Sharma, K. (2014). Inpatriation as a key component of global talent management. In *Global talent management* (pp. 79–92). Springer International Publishing.
29. Reiche, B. S. (2011). Knowledge transfer in multinationals: The role of inpatriates' boundary spanning. *Human Resource Management, 50*(3), 365–389.

30. Tahvanainen, M., Welch, D., & Worm, V. (2005). Implications of short-term international assignments. *European Management Journal, 23*(6), 663–673.
31. Tansley, C. (2011). What do we mean by the term 'talent' in talent management? *Industrial and Commercial Training, 43*(5), 266–274.

The Transferable Skills Development Programme of a Portuguese Economics and Management Faculty: The Perceptions of Graduate Students

Iris Barbosa, Carla Freire and Mariana Paiva Santos

Abstract The Bologna Declaration imposed significant changes in European higher education, placing emphasis on teaching methods that promote students' active learning and the development of skills, including both the skills that are specific to the academic qualification area and the ones that became known as *transferable skills*. The latter have been considered to be fundamental to the promotion of graduate employability. This chapter presents the specific case of the *Transferable Skills Development Programme* of a Portuguese public university and analyses the perceptions of 21 graduate students, in the area of Economics and Management, regarding its relevance and contribution to the development of transferable skills. Through the completion of qualitative questionnaires and the interaction established within three focus groups, the respondents acknowledged the innovative profile of the programme in the Portuguese university context and its contribution to the development of transferable competencies, which they believed to be especially relevant to their future job performance. The graduate students involved considered that the *Transferable Skills Development Programme* should continue to reinforce its commitment to activities and contents which were markedly more practical.

1 Introduction

Significant changes in the labour market, such as the failure of the Taylorist/Fordist model, the globalization of the economy, as well as the emergence of the so-called knowledge-based economies, dictated the higher value attributed to new qualifications and skills in the workplace. During this process, the management of people has moved away from a focus on function and has centred on the individual

I. Barbosa (✉) · C. Freire · M.P. Santos
University of Minho, School of Economics and Management, Braga, Portugal
e-mail: iris@eeg.uminho.pt

C. Machado (ed.), *Competencies and (Global) Talent Management*,
Management and Industrial Engineering, DOI 10.1007/978-3-319-53400-8_2

himself, on his flexibility and capacity to adapt to change [1]. Literature suggests that it is no longer sufficient for a worker to be able to perform specialized technical functions; it is essential that he is also capable of demonstrating a wide range of skills which are required in the most diverse professional contexts, i.e. transferable skills [2–7].

Above all, the issues related to graduate employability that have dominated concerns in contemporary European higher education [4, 8–13] highlighted the relevance of including competencies in the curricula, which are expected to promote the new graduate's access to and success in the job market. The Bologna Declaration [14] has established a turning point in these debates, demanding that European universities adopt a new teaching-learning paradigm. This favours active student-centred learning processes and is grounded on the development of skills for each area of knowledge, both technical and scientific, as well as transferable skills.

Various European universities have sought to respond to these challenges through the promotion of a broader spectrum of education. This is centred on activities that foster the skills and attitudes which are valued and required by the employers [15–18], thus contributing to complement students' education and training and positively distinguishing graduates in the labour market. Some examples of universities that have incorporated programmes with these objectives are the Universities of Cambridge[1] and Edinburgh[2] (United Kingdom), the University of Freiburg[3] (Germany), the University of Limerick[4] (Ireland), the University of Utrecht[5] (Holland), and the University of Zurich[6] (Switzerland). In Portugal, one can highlight the Nova School of Business and Economics,[7] the ISCTE,[8] as well as the University of Minho.[9]

This chapter presents the case of the *Transferable Skills Development Programme* at the Economics and Management faculty of a public university in the north of Portugal. It further analyses the perceptions of graduate students concerning the relevance of transferable skills in their sound job performance, as well as the contribution of this particular programme to the acquisition and development of these very skills. Since this investigation constitutes a pioneer study of this specific programme, the results could contribute to its improved adjustment to

---

[1]http://www.skills.cam.ac.uk/undergrads/skills/.

[2]http://www.ed.ac.uk/schools-departments/institute-academic-development/postgraduate/taught/courses-events/open-workshops.

[3]http://www.studium.uni-freiburg.de/studium-en/bok-en.

[4]http://www2.ul.ie/web/WWW/Services/Research/Graduate_School/Current_Students/Training_for_Research_Students.

[5]http://www.uu.nl/EN/informationfor/intstaffandvisitors/Research/policyprojects/phace/Workshops/Pages/Useyourtransferableskills.aspx.

[6]http://www.ueberfachliche-kompetenzen.uzh.ch/index_en.html.

[7]http://www.novasbe.unl.pt/php/templates/article_simple.php?id=34.

[8]http://ibs.iscte.pt//?pt=programa-de-desenvolvimento-pessoal.

[9]http://www.eegs.eeg.uminho.pt/.

students' expectations and needs, as well as to the development and/or improvement of programmes with similar objectives in other university contexts, both on a national and international level.

2 Theoretical Background

The concept of competence stems from Taylor's "Principles of Scientific Management" in 1911 [19]. Through the scientific analysis of work, and particularly the observation of the time and motion of the most competent workers, Taylor was committed to identify the one best way to complete each task and to transform it into a patter that should be observed by every worker. The objective was to guarantee worker competence and, consequently, to improve the organizational efficiency [19, 20].

Until the 1970s, corporate management was dominated by the logic of continuous mass production, supported by the manual control of manufacturing through the use of the called "craft" skills. This resulted in great specialization, division of labour into small tasks, and control management hierarchy [5]. From this decade onwards, workforce globalization and customer orientation have created the need for companies to rely on multifaceted workers, who are able to adapt quickly to market demands. The production cycle has become shorter, and the responsibility for the manufacturing process has been transferred to the workers themselves [5]. These changes in the organization's operation scheme have led to the fact that people management is no longer centred on function but rather on the individual himself, on his flexibility and ability to adapt to change [1]. In this context, the personal skills and attitudes that will help individuals to adapt to changes and to the new demands in the workplace have assumed great importance [21].

2.1 The Concept of "Competence" and Different Approaches

McClelland [22] was a pioneer author in the definition of the concept "*competence*", in his famous article "*Testing for competence rather than intelligence*". He believed that the traditional selection tests presented several shortcomings; namely, they did not allow for the measurement of all the aspects which are relevant to the performance of an established function. Academic certifications and intelligence tests are not in themselves wholly sufficient in predicting an employee's future performance; it is for this reason that it would be necessary to adopt a more comprehensive concept, namely that of competence.

The term "competence" has since been defined differently in literature, by researchers in the various knowledge areas [23]. Psychology refers to the concept as a measure of ability, associating performance to personality traits and underlying capacities. Management uses a functional analysis of competencies to determine how the improvement of individual performance contributes to the attainment of an organization's objectives. sees competence as a key concept for the implementation of a strategic direction in processes such as recruitment, selection, training, performance assessment, promotion, reward systems, and staff planning. Finally, education relates the concept to the idea of preparing for employment and as professional acknowledgement. As a result of this fragmentation, literature in this area is divided into two approaches [5, 19, 24, 25]. These use specific terms and give them distinct meanings: (a) *competence*: understood as a performance standard, i.e. which allows for the efficient performance of a task and (b) *competency*: considered as the behaviour one must reveal in order to carry out the job tasks and functions competently.

In the UK, the concept of competence was associated to the *Management Charter Initiative* (MCI). This undertook an analysis of 3000 managers and culminated in the production of a typology, which established the competencies required for managers in various organizations and in diverse sectors. This initiative was used as a reference for the drafting of the NVQ (*National Vocational Qualification*) in 1988. This defines "competence" as the capacity to perform a specific function, perceived as a performance standard, which allows one to acknowledge and qualify a job as being efficient [24, 26].

In the North American context, the approaches tend to focus on the individual and on the behaviour he must reveal to ensure competent performance [24]. Studies which subscribe to this view were mainly carried out in the context of education/training and deal with the acquisition of tools for more efficient task performance [26].

There is, in addition to the above, another approach to be found in literature which focuses on the person's qualities and characteristics (*competencies*) as constituting determining factors for higher performance [24, 27, 28]. Levy-Leboyer [29] relates competencies to an individual's psychological aspects, such as personality traits, skills, and the acquisition of knowledge. Since the latter approach constitutes a comprehensive perspective of the concept of competence [19, 24, 28, 30], it includes all three aspects: requirements or qualities needed for a specific task (*competencies*), observed behaviour (*competency*), and result of an individual's performance (*competence*).

Woodruffe [25] classifies competencies in two nuclei: (a) technical skills which are specific to the job and (b) generic skills, which can be universal or transferable. In the UK, the DfEE (*Department for Education and Employment*) and the *Association of Graduate Recruiters* (AGR) proposed a classification of competencies [7], which is presented in Table 1.

Table 1 Classification of competencies

Competency title	Definition
Key (or core) skills	The general skills required for various jobs, which include basic literacy and numeracy, as well as a set of transferable personal skills such as the ability to work well with others, communication, self-motivation, and the ability to organize one's own work and to use information technology
Vocational skills	The skills required for specific jobs or occupation groups, which are less useful outside these areas. They are less general than *key skills*, but can be transferred from a job in a specific area to another
Job-specific skills	Skills which are specific to a particular job or even an organization, also designated as technical skills

Source Adapted from Stewart and Knowles [7]

Lawrence [31] proposes a typology which has been adopted in the USA:

(i) Academic skills (knowledge and skills associated with the academic disciplines of reading, writing, mathematics, and science);
(ii) Employability skills (used to perform effectively, which are transferable to a broad range of occupations, such as teamwork, decision-making, and problem-solving); and
(iii) Occupational and technical skills (specific technical and occupational knowledge and skills which are job-specific, such as knowledge of sales methods, engine repair, and database programming).

The concept of "key (or core) skills" [7] thus draws closer to "employability skills" [31], in the sense that both refer to skills which are transferable to different subject areas and professional contexts. The following section analyses this concept in greater detail, from now on designated in this chapter as "transferable skills".

2.2 *Transferable Skills: Definition of the Concept*

As is the case for a wide-ranging concept of competency, there is also no consensus as to the concept of transferable skills. Various typologies have been proposed, and designations for the concept have been put forward. Tien et al. [32] have presented some examples of these designations and have situated them geographically, namely "*employability skills*" (*National Skills Standard Board*, USA); "*core skills*" (United Nations); "*key competencies*" (Australia); "*core skills/key skills*" (Great Britain); "*employability skills*" (Canada); and "*basic competencies*" (Taiwan). Other designations for the concept are, for example, "*Transferable Skills*" (*Training Agency*, UK) and "Common Skills" (*Business and Technology Education Council,* UK).

According to Mansfield [5], transferable skills are all the competencies which are not technical, specific, or occupational. For Assiter [2], transferable skills are the generic capacities which allow individuals to attain success in a wide variety of tasks and occupations. The *Department for Education and Employment* (DfEE), in the UK, considers transferable skills to be those competencies which are essential for performance in all sectors and at all levels [3]. Drummond et al. [4] mention these as skills which can be transferred to contexts outside the academic field of study, and Gibbons-Wood and Lange [33] describe them as being those which support competent behaviour in all areas.

Harvey et al. [12] suggest two categories of transferable skills: (i) personal attributes, which comprise knowledge, continuous learning, flexibility and adaptability, self-regulation, self-motivation, and self-confidence and (ii) interactive attributes, which encompass communication, relationships, group work, and the ability to influence.

At Luton University in the UK, the study carried out by Atlay and Harris [34], which was the result of the communication established among this university, its employer partners, and the community, subscribes to a competency model that is divided into four large areas: data collection and processing, communication and presentation, planning and problem-solving, as well as social development and interaction.

In Australia, a study undertaken by Kearns [35] separates transferable skills into four large groups: (i) preparation for employment and working habits; (ii) interpersonal skills (supported by personal attributes and values, such as emotional intelligence and the understanding of oneself); (iii) entrepreneurship, innovation, and creativity; and (iv) skills related to learning, thinking, and adaptability. This study also emphasizes the competency model adopted by the United Nations, which contemplates three categories: core skills (communication, teamwork, planning and organization, responsibility, creativity, customer orientation, commitment to continuous learning, and technological awareness); core values (integrity, professionalism, and respect for diversity); and management skills (leadership, vision, the development of others, confidence building, performance management, decision-making, and judgement).

In the context of the European project *Tuning*, whose objective was to establish an exchange of information and collaboration in the development of quality, efficiency, and transparency in more than 100 European universities, both generic and specific skills were analysed and associated to different areas of study. In the report for the first phase of this project, the authors González and Wagenaar [36] classified general skills into 3 groups: (i) instrumental skills (cognitive, methodological, technological, and linguistic abilities); (ii) interpersonal skills (social interaction and cooperation, and critical and ethical consciousness); and (iii) systemic skills (the ability to analyse the whole and understand how the parts work together, as well as how to combine and apply skills and knowledge to different situations), with the last of these groups requiring the prior acquisition of skills from the first two. This project highlights the most important skills as being: the ability to analyse and synthesize; the ability to learn, solve problems, and apply knowledge to practice; the ability to plan and organize, to work autonomously, and to adapt to new situations; the ability to establish interpersonal relationships and work in a team, to communicate orally and in writing (in a native language and in a second language); and the ability to manage information and, finally, the concern with quality.

2.3 The Development of Transferable Skills in the Academic Context

The Bologna Declaration [14] produced a paradigm shift in the teaching-learning process of European higher education. This has become more student-centred and has focused on student's ability to obtain knowledge and acquire new skills. It also advocates the increasing importance of transferable skills in graduates' employability [36, 37], which ensues from the belief that technical skills are insufficient in the context of the present model of work organization and given the current competition and demands of the labour market. Literature, however, reveals that there is a gap between the skills developed at universities and those which are required in the workplace [38, 39], precisely pointing to the need for new skills, as well as to the complementarity of transferable and technical skills.

Various studies have focused on the British academic context [10, 13, 40–45], where there seems to be a predominant perception that graduates are rather badly prepared for the conditions they will be confronted with in the labour market [8]. For example, the purpose of the development of the initiative "*Enterprise in Higher Education*" (EHE) in 1987 was to surpass these shortcomings in training. Binks [8] mentions that the two reasons for implementing this initiative were that: (i) students are not often exposed to work situations during the course of their academic careers and (ii) the curriculum is designed in such a way that it does not encourage students to acquire other skills. This initiative led to the introduction of new teaching-learning strategies in various higher education institutions, which aimed to improve students' ability in the transferable skills required by the labour market [42]. An example of these experiences is that of the University of Nottingham, which implemented a systemic approach to analyse training needs, with a view to the constant improvement of learning, teaching, and assessment [8].

On the basis of a review of literature, Drummond et al. [4] briefly presented three broad approaches that can be implemented by higher education institutions to develop skills within the curriculum:

1. *Embedded or integrated development*: the development of skills in an integrated manner within the curriculum; this can occur at different levels of the programme (randomly interspersed, core modules, mapped skills with/with no progression, and project-based development);
2. *Parallel* (*stand-alone*) *development:* the development of skills in free-standing modules, which are not integrated into the curriculum; and

3. *Work placements or in work-based projects*: the student spends a period of time in practice, which is the case of *sandwich* courses. These consist of a theoretical component and of a practical one, with the integration of time periods spent in a professional context, which allow for experience in the real-life context of work. This model is considered by employers as the best way to develop students' employment-related skills.

2.4 The Development of New Skills from the Students' Perspective

Several studies have sought to analyse the perceptions of students in higher education regarding the acquisition of transferable skills during their academic path, as well as when they begin their professional lives [12, 13, 41, 43, 44]. The study carried out by Nabi and Bagley [13] evaluated graduates' perceptions in relation to the importance and quality of the transferable skills acquired during their academic path. It focused on students' satisfaction with the course itself, their university experience, as well as preparation for their future career. This study revealed that graduates tend to rate the importance of skills highly than their own ability in those. The results also presented employers' views as to the quality of transferable skills, indicating that these should be improved and adjusted to the needs of the labour market. Identical conclusions were reached by other studies, which reveal dissatisfaction in relation to the skills acquired by students in the academic context [40].

The study by Rosenberg et al. [46], in the North American context, examined the transferable skills required for job performance, the way in which these are developed during the course of an academic career, and the need for additional training after concluding a degree. This study was grounded on the perceptions of three distinct groups: recent graduates, the faculty that taught them, as well as the human resource managers who recruit them. More specifically, the participants in this study responded to a survey that included 47 items measuring eight dimensions of transferable skills: basic literacy and numeracy, critical thinking, management, leadership, interpersonal, information and communication technologies (ICTs), systems thinking, and work ethic disposition. This study confirmed the importance of transferable skills and of the communication across students, the university, and employers in the development of these skills.

The study conducted by Whittle and Eaton [45] analysed the introduction of a transferable skills development module in the first year of a degree in medicine at a British university, as well as its repercussion throughout the course. Students perceived that it has enhanced their self-confidence during the educational process and contributed to develop their skills in most areas, particularly with regard to self-learning skills. The results suggest that students feel they are better prepared to succeed in a learning system that grants them greater responsibility and independence for their own development of skills.

Despite the fact that most studies attribute great importance to transferable skills, there are studies which tend to mitigate it. For example, the study developed by Laughton and Montanheiro [43], which sought to assess the perceptions of recent graduates who were integrated in professional life, concluded that the latter gave little importance to transferable skills when applying for a first job.

In the Portuguese university context, the research conducted by Cabral-Cardoso et al. [47] sought to determine the extent of the importance of transferable skills to the graduates' competent job performance, as well as the contribution of higher education institutions and employment entities to their acquisition and development. The study started with a list of 40 transferable skills which were identified by the authors as being dominant in literature. These are indicated in Table 2.

The study by Cabral-Cardoso et al. [47] concluded that there is a consensus between students and employers as to the transferable skills which are considered to be the most important in the labour market. Both groups attributed great value to instrumental skills, such as work planning and organization, problem-solving, information and communication technologies, personal relationships, and continuous learning. However, it is the employers who tend to evaluate more favourably the importance of transferable skills in everyday work, while students tend to underestimate them and consider technical skills to be essential. In addition, it is the employers who tend to attribute greater relevance to the following transferable skills: numeracy, foreign languages, business sensitivity, capacity for teamwork, and conflict management.

Table 2 List of the most common transferable skills found in the literature

ICT	Innovation/creativity	General knowledge	Initiative
Oral communication	Leadership	Continuous learning	Persistence
Written communication	Information collection and processing	Attention to detail	Self-control
Teamwork	Planning and organization	Influence/persuasion	Decision-making
Customer orientation	Intercultural competence	Questioning skills	Motivation/ personal drive
Problem-solving	Critical thinking	Listening skills	Conflict management
Numeracy	Ethics	Interpersonal relationships	Motivating
Foreign languages	Business sensitivity	Action planning	Networking
Autonomy	Stress tolerance	Negotiation	Risk-taking
Adaptability	Self-confidence	Self-presentation	Developing others

Source Based on Cabral-Cardoso et al. [47]

3 The Case of *the Transferable Skills Development Programme* in a Portuguese faculty of Economics and Management

The Bologna Declaration significantly changed the European higher educational system, particularly by placing emphasis on student's role and favouring active learning based on the development of skills, both of those specific to each academic qualification area and of transferable skills. These changes stimulated the development of a *Transferable Skills Development Programme* in the faculty of Economics and Management of a public university located in the north of Portugal. The aim of this programme, which was implemented in the academic years of 2012/2013, was to complement the development of the transferable skills already included in the course syllabus through a range of activities and initiatives. It also proposed to enhance students' awareness of the importance of these skills as a distinguishing factor, which generates opportunities of success in the labour market.

The programme has been groundbreaking in the context of this Portuguese public university, which distinguishes itself from others by of its own regulations, structure, and operation. In its implementation year (2012/2013), the programme was directed to Master's degree students, a fact which coincided with a curricular revision for these courses. In fifteen of those master courses, a 7.5 ECTS curricular unit was included; this consisted of a Methodology/Project module (4.5 ECTS) and a Transferable Skills module (3 ECTS). The latter was related to participation in activities of the *Transferable Skills Development Programme*. At present, this programme is also available for Honours and Doctorate degree students, although priority is given to the Master's degree students due to the integration of these activities into the regular study plan and their corresponding ECTS credits.

The activities and initiatives developed within the framework of this programme specifically aim to promote the acquisition and development of ten skills, which have been deemed to be relevant for students graduating from the Economics and Management area. During the first phase, the identification of these skills resulted from a revision of literature [47, 48], and in the subsequent phase, it ensued from the assessment undertaken by the employers who maintain regular contact with this faculty and collaborate on several of its activities (e.g. workshops, lectures, and field days). The following skills, which are included in the programme, emerged from this analysis: (1) interpersonal relationships, (2) teamwork; (3) leadership, (4) communication, (5) creativity and innovation, (6) ethical awareness and critical thinking, (7) planning and organization, (8) ICT, (9) problem-solving, and (10) results orientation.

The activities developed by the *Transferable Skills Development Programme* throughout the academic year are based on two typologies, which presuppose different levels of participation by the students: (1) courses and workshops and (2) lectures and events. The "courses" consist of activities developed over various sessions, which are related to useful work tools in the academic context, as well as in professional life, with a maximum duration period of 12 h (e.g. SPSS, Excel,

Stata, and NVivo). The "workshops" comprise single sessions of presentation and discussion of practical content, followed by activities and exercises performed by students. These workshops deal with themes such as job-seeking, career development, leadership, interpersonal relationships, problem-solving, decision-making, organization, self-development, and research techniques. These activities presuppose the active participation of students and are limited to a maximum number of participants (between 20 and 50). In the "lectures", a guest speaker (e.g. CEOs from reference companies, former students from this faculty in prominent positions in the labour market) presents his life story and that of his professional experience, which lasts from 1 to 3 h. The "events" involve the organization of events around a central theme (e.g. a series of lectures, round-table discussions, and parallel sessions), which can last from a few hours to the entire day. The number of participants in these last activities is higher and is only conditioned by the seating capacity of the auditorium or the premises used. Within this framework, there are also study trips to companies in various sectors of activity, which are an opportunity for students to obtain some knowledge of the everyday running of companies. Included in the programme are also large events such as employability day (the Job fair), research day, and job sessions (days of training provided by companies)

In 2015/2016, this *Transferable Skills Development Programme* was in its fourth operative year. During this year, there was an increase in student participation, adding up to a sum total of 1476 students (920 participants from the 1st cycle, 487 participants from the 2nd cycle, and 69 participants from the 3rd cycle), which refers to 6424 effective acts of participation in the 129 organized activities. One must emphasize that current programme feedback results from the assessment of each of the activities, which is done on paper by the participating students. The intervening parties who collaborate in the various activities (e.g. speakers and trainers) supply informal feedback about these, and thus, there is no duly recorded written assessment.

4 The Research Design and Sample

The purpose of this study is to get to know the perceptions of graduate students attending the faculty of Economics and Management of a public university in the north of Portugal, regarding the importance of the transferable skills to their present and/or future job performance. It specifically aims to analyse their opinions concerning the contribution of the recently created *Transferable Skills Development Programme*. Due to the fact that there are very few similar studies in Portugal [47], and to the absence of studies addressing this specific programme, the present study is of an exploratory nature. Therefore, we considered suitable to adopt a qualitative method and resorted to two data collection instruments. In the first phase, graduate students were asked to complete a qualitative questionnaire with the objective of gathering basic information about the topic. This questionnaire consisted of the following three questions:

1. Which are the ten transferable skills that you consider to be the most important?
2. How do you evaluate the *Transferable Skills Development Programme*?
3. Which suggestions of improvement would you like to present to the *Transferable Skills Development Programme*?

Subsequently, the same graduates were divided into three focus groups. This number was regarded as sufficient, since the study contemplates a very particular population [49]. This technique has been considered propitious to stimulate the individuals' interactive discussion of ideas regarding each topic [50]. These objectives are particularly relevant when the theme has been minimally explored. Five questions were presented to each focus group with the purpose of generating discussion among the participants. The order of presentation varied, according to the dynamics established:

1. Which are the ten transferable skills you consider to be the most important and why?
2. What is the relevance of the transferable skills, when compared to the technical skills, for the competent job performance of young graduates and why?
3. How do you evaluate your ability in the transferable skills that you have identified as especially relevant?
4. Which factors contributed/have contributed to the development of these transferable skills?
5. How do you evaluate the role performed by the organization you are employed by in your acquisition/development of these transferable skills?
6. How do you evaluate the performance of the *Transferable Skills Development Programme* in your acquisition/development of these transferable skills?

Twenty-one graduate students from five different Master's degree courses participated in this phase. Each group consisted of seven students, a number that has been considered adequate [49, 51]. The sample shares two features that are deemed fundamental to the objectives of this study: (1) the individual has already benefitted from the *Transferable Skills Development Programme* and (2) the individual has already had some relevant professional experience. The existence of common core features shared by the participants, thus making the focus groups more homogenous, is considered fundamental when aiming to capitalize on the experience shared on a given topic [52], as well as to generate relaxed and open conversations [53].

The sample consisted of 16 Portuguese women and 5 men, one of these was Portuguese and the four others were Brazilian. They were, therefore, all Portuguese native speakers. The youngest respondent was 23 years old, and the oldest was 41; the average age was approximately 31 years. All individuals were employed at the time of the research. The fact that some of the graduate students already knew each other was evaluated as propitious to a greater interaction and debate of ideas [49]. Each focus group meeting lasted approximately 1 h 15 min and was audio-recorded. Several field notes were collected by one of the researchers during the focus group meetings.

The following chapter presents the main results of this research. Emphasis is placed on the results revealing shared opinions by the graduate students, as well as on more peculiar viewpoints. Fictional names were used to protect the participants' anonymity.

5 Findings

5.1 The Importance of Transferable Skills Versus Technical Skills

The study participants assessed the technical skills as indispensable to competent job performance, yet they believed it was important to conjugate these with transferable skills. They thus evaluate them as complementary, and such has been advocated by the literature [30, 54, 55]. These results are different from those obtained by Cabral-Cardoso et al. [47] a decade ago, when it was found that students tended to underestimate transferable skills and to consider technical skills as being more essential. The following excerpts illustrate this current perception of the complementary relationship between transferable and technical skills:

> Samuel (27 years): *They go hand-in-hand. As you acquire more technical competence, you will get to develop your technical skills better.*

> Bruno (35 years): *You must have deeply-rooted technical skills to be able to develop your transferable skills. One thing complements the other, but each one on its own is not really valued. You must have them together.*

Some participants commented that the real understanding of the importance of transferable skills only occurs during professional experience. This was expressed by Helena (32 years): "The professional experience I now possess is what showed me the importance of having transferable skills too". This result corresponds with the study undertaken in New Zealand by Rainsbury et al. [55], which revealed that whereas business graduates attribute equal importance to transferable and technical skills, business students without professional experience assess transferable skills as being less relevant. During the discussion of this topic, a participant exemplified the important use of transferable skills when exercising her professional activity:

> Helena (32 years): *More than giving the customer an impression of the technical skills I possess, it will be my use of transferable skills that will convince him to buy.*

Several respondents revealed the perception that while technical skills are expected and required at any organizational level, mastering transferable skills is especially relevant in higher job positions and crucial in building a career:

> Maria (27 years): *I agree with that view [the great relevance of transferable skills] when talking about a higher job position.*

> Lucas (34 years): *They [in the organisation] focus on technical skills when we're in lower job positions but if you want to climb the ladder, then transferable skills are essential.*

The results therefore suggest that the students without professional experience at this faculty of Economics and Management may be less sensitive to the great importance of transferable skills in the current corporate world [43, 56]. This aspect could limit their participation in initiatives developed by this institution, namely those of the *Transferable Skills Development Programme*.

5.2 The Most Relevant Transferable Skills

Through the completion of the qualitative questionnaire and during the focus group meetings, the research participants chose and discussed which transferable skills they believed to be the most important to their current and/or future professional activity, from the list considered by Cabral-Cardoso et al. [47] in their research also conducted in the Portuguese context (represented in Table 2). Table 3 summarizes the graduate students' opinions and presents illustrative quotes.

These results are rather different from those found 10 years ago by Cabral-Cardoso et al. [47], when working with students and graduates from the north of Portugal. It coincided only in the relevance attributed to planning and organization skills, as well as motivation/personal drive. On the other hand, the

Table 3 The ten most relevant transferable skills according to the study participants

Transferable Skills	Illustrative quotes
Planning and organization	Eva (32 years): T*he skills that are lacking when people leave university, such as how to organise, knowing how to plan, being able to solve difficult problems…*
Problem-solving	Helena (32 years): *One of the skills I find important to develop is that of problem-solving*
Adaptability	Lucas (34 years): *Having the capacity to adapt… For example, in my career we are always changing from one place to another*
Creativity and innovation	Bruno (35 years): *Creativity is very important! But, at the same time, it is one of the skills which is most difficult to develop*
ICT	Maria (27 years): *Mastering information technology is a key selection criterion to my organization*
Leadership	Lucas (34 years): *All military personnel must have this skill [leadership]*
Teamwork	Samuel (27 years): *Teams must be able to work towards the same goals*
Oral communication	Diana (35 years): *I highlight oral communication: if I can't understand and adjust my discourse to who is on the other side, I may lose out on the business*
Interpersonal relationships	Monica (30 years): *It's important to know how to work with different people: people from different contexts, with particular experiences and specific problems*
Motivation/personal drive	Maria (27 years): *Being motivated helps you to deal with frustration*!

importance of oral communication and interpersonal skills expressed by participants in this current study was also detected in the study carried out by Andrews and Higson [18], which covered four European countries (UK, Austria, Slovenia, and Romania). In the case of ICT and teamwork skills, these were already evaluated as being especially pertinent by graduate and undergraduate students further afield in New Zealand [55]. Various other studies point to an identical perception of the importance of skills in the areas of planning, organization, oral communication [11, 13, 40], adaptability [57], leadership [6, 46], motivation/personal drive [7], as well as creativity and innovation [40]. The results indicate that there has been some development in comparison with the data collected by Cabral-Cardoso et al. [47]. Moreover, they show greater proximity between the perceptions of students in Portugal and those of other countries regarding the identification of the transferable skills that are deemed as particularly important to competent job performance. This result does not seem to be extraneous to the increasing similarity of the structures and curricular programmes of higher education [58], as well as to the complex needs imposed by the increasingly global business world.

From the list of transferable skills presented in Table 3, participants elected the following as being those which they already reveal significant development: problem-solving, adaptability, and motivation/personal drive. On the other hand, they admitted a weaker development in the skills related to interpersonal relationships. The respondents also considered that some of these transferable skills are easier to acquire/develop through formal mechanisms and programmes (e.g. ICT) than others (e.g. creativity and innovation).

5.3 *Mechanisms and Programmes for Developing Transferable Skills*

Graduates expressed the view that the development of transferable skills is a continuous and permanently unfinished process, whose construction ensues from a variety of experiences and contexts. The following comment illustrates this statement:

> Olivia (29 years): *All these skills can be continuously developed. I don't think someone gets here and says: "I'm at this level, I've finished, and it's over". By other words, everything can be developed.*

Various participants expressed the belief that the employer should invest in training programmes and actions to develop collaborator's transferable skills, specifically the ones that are markedly relevant to the individual's performance and, consequently, to the organization's overall strategy and success. Examples were provided regarding training on customer service skills:

> Maria (27 years, employed):. *The organization provides training on technical and transferable skills. [For example,] in emotional and conflict management, which is very important in the case of permanent customer service.*

Teresa (26 years, employed): *The hospital where I work also provides compulsory training in transferable skills, which is the case of communication with patients. This is compulsory for all professional categories.*

Some of the respondents considered, however, that the lack of investment in transferable skills development programmes by employers constitutes no excuse for the employee to stagnate in this area. On the contrary, they referred that employees should play an active role in seeking the best means and mechanisms to develop transferable skills continuously, namely by resorting to external training:

Julia (29 years): *The skills which are missing… I try to respond with training, for example in technologies, in languages…*

Diana (35 years): *Same opinion here: although I have no access to training at work, I can also look for training outside.*

Lucas (34 years): *I agree with you. You must have personal initiative (...) if you're interested, if you have the initiative, you end up learning! If you don't, you're going to stagnate in your career.*

Furthermore, the participants in this study listed other mechanisms and sources which they believe to contribute to the acquisition and development of transferable skills. The replies underline the importance of the family context (useful in the acquisition of moral values), social interaction (e.g. with school or work colleagues), hobbies (e.g. the development of teamworking skills, resilience, and tolerance of stress through sporting activities), volunteer work, as well as other activities in the academic context (e.g. participation in student associations). The individuals added to this the development of transferable skills which ensued from any experience in the world of work, even if this was during a short period of time and very different from their academic qualification area (e.g. the development of customer service skills). This result is in line with the "discourse of experience" [59], which portrays the belief by students that any experience in the organizational environment contributes to the acquisition of transferable skills and positively differentiates the student from his colleagues. Two participants emphasized the conjugation of various experiences as being favourable to the development of transferable skills:

Maria (27 years): *We are the sum total of everything around us, family, school, hobbies…*

Lucas (34 years): *Totally agree: Transferable skills are the result of experiences accumulated throughout life.*

5.4 *Perceptions of the Transferable Skills Development Programme*

The study participants highlighted the importance of the university's role in providing programmes and tools for the acquisition and development of transferable skills, a position which is in accordance with what is found in the literature [60, 61].

In this regard, the participants acknowledged the positive contribution of the *Transferable Skills Development Programme*, and underlined its "innovative and distinctive profile" in the context of Economics and Management faculties in Portugal, as well as its "sound structure and wide-ranging programme", "excellent content", and "enriching lectures". This positive view seems to be associated with the great correspondence between the skills developed in the programme and those which the study participants consider to be the most relevant, as presented in Table 4.

In addition to this generally positive assessment, the participants highlighted the great usefulness of the *Transferable Skills Development Programme* for students without working experience. The following comment illustrates this opinion:

> Monica (30 years): *It's very useful. The Transferable Skills Development Programme allows you to address specific needs which the courses, be they Master's or Honours, are unable to do, namely in the development of transferable skills. It contributes to students feeling more confident in the world of work.*

Various suggestions of improvement of the *Transferable Skills Development Programme* were then presented. These focused on a better adjustment of programme time schedules for employed students, as was explained by the following respondent:

Table 4 The transferable skills developed in the Transferable skills development programme versus the transferable skills valued by the study participants

Transferable skills	Transferable skills development programme	Valued by the study participants
Planning and organization	✓	✓
Results orientation	✓	
Problem-solving	✓	✓
Adaptability		✓
Critical thinking	✓	
Creativity and innovation	✓	✓
ICT	✓	✓
Leadership	✓	✓
Teamwork	✓	✓
Oral communication	✓	✓
Interpersonal relationships	✓	✓
Motivation/personal drive		✓
Ethical awareness	✓	

Teresa (26 years): *You should direct the programme to the after-work time period as well. Our choices [from the available content] end up being rather restricted and are made, above all, because of our respective time schedules.*

Most of the participants further consider that the *Transferable Skills Development Programme* should continue to reinforce its commitment to clearly practical activities and contents. The following comment reveals the dissatisfaction ensuing from content evaluated as being more theoretical and abstract:

Julia (29 years): *I did the conflict management course. But I can say that, basically, it dealt with theoretical psychological concepts, which were difficult to implement in practice.*

A couple of participants also questioned the selection of some of the topics covered in the *Transferable Skills Development Programme* and considered them to be less suitable for their work objectives, as the following comments reveal:

Angela (24 years): *The programme also contemplates skills which are important when carrying out research, contents which are more theoretical and are not so important for companies. But most of us don't want to do research; we really want to get into the business world.*

The conversation then moved to the benefits of a curricular traineeship programme, with several students concluding that the *Transferable Skills Development Programme* does not act as a replacement for it. Various respondents showed the preference for a traineeship period at a company:

Olivia (29 years): *The shortcoming of our Master's is that there is no traineeship phase. The Transferable Skills Development Programme is, supposedly, for us to have useful skills in the work context. But wouldn't a period of traineeship be far more effective and efficient in getting to develop those skills?*

Monica (30 years): *A [better] way of acquiring these skills would be on the basis of work, on the basis of traineeship, even if it only lasted a month or two.*

This perception on the part of the graduate students that contact with the "real-world problems" by means of a traineeship phase is fundamental to the development of transferable skills was also identified on the studies conducted by Andrews and Higson [18] and Crebert el al. [62], namely with regard to leadership, oral communication, and interpersonal relationships skills.

In summary, despite the innovative characteristics of the *Transferable Skills Development Programme* in the Portuguese university context, and its outstanding contribution to the development of transferable skills among its graduate and undergraduate community, this research points to the need for greater adjustment to students' requirements and expectations. The graduate students who participated in the study presented several suggestions which may contribute to the improvement of this specific programme, as well as to be taken into consideration by other faculties/universities when designing programmes and tools with similar mission and goals.

6 Conclusions

The Bologna Declaration imposed significant changes in European higher education, placing emphasis on teaching methods that promote students' active learning as well as on the development of skills, whether specific to the academic qualification area or to what became known as *transferable skills*. The latter have been considered to be fundamental to the promotion of graduate employability. This research sought to learn the perceptions of graduate students, from the faculty of Economics and Management of a public university in the north of Portugal, regarding the importance of acquiring and mastering transferable skills and, especially, in relation to the contribution of the recently created *Transferable Skills Development Programme* in this regard. Results point to the great importance attributed by graduate students to transferable skills so as to promote great job performance and a successful career. Of the ten skills classified as the most relevant, graduates included professional skills (e.g. planning and organization, problem-solving), communication skills, interpersonal skills, and personal skills (e.g. adaptability and motivation/personal drive). The current activities of the *Transferable Skills Development Programme* focus on the general development of these transferable skills.

During the three focus group meetings, a view prevailed that the programme should reinforce its commitment to practical activities which promote the development of skills that meet the needs of the corporate world. Additionally, various participants advocated their preference for a curricular traineeship programme, considering it a particularly efficient means for developing such transferable skills. These results point to the need for the *Transferable Skills Development Programme* to especially focus on activities of an eminently practical nature, such as workshops of a longer duration and/or which are undertaken in an organizational context. The results also suggest a need to increase the number of visits to companies, as well as of other activities providing opportunities for closer proximity to those in leadership positions. Despite the diversity of activities and their time schedules, the individuals indicated the need for a better adjustment of the schedule of activities to the availability of working students, advocating the development of more activities in the time periods after work and at the weekend.

What seems to be particularly relevant is the graduate students' belief that the awareness of the importance of transferable skills for and successful career-building only really occurs after acquiring some relevant work experience. This result highlights the need for those responsible for the programme, as well as for the supervisors of the various Honours and Master's degrees at this faculty of Economics and Management, to generate greater awareness among students for the importance of the transferable skills. Some of the lectures promoted by the *Transferable Skills Development Programme* seem to be excellent platforms for the reinforcement of this message, namely the ones given by speakers who share their life stories and their professional paths—such as former graduates from this faculty of Economics and Management that occupy prominent positions in the market, as well as CEOs from companies of reference. Finally, this study reveals that the *Transferable Skills Development Programme* must be subjected to an efficient and continuous system of evaluation by graduate and undergraduate students. This should include room for the presentation of suggestions of improvement.

Although this research study contemplated a specific programme and academic context, the emerging results and recommendations could be equally relevant in the development and/or adjustment of any academic programme, within any Portuguese or European academic framework, aiming to set up a first-rate tool for the development of transferable skills among students and graduates. This would complement the often inadequate role of Honours and Master's degree course plans in this regard.

There are some limitations in this study which deserve special reference, since they provide clues for future investigation work. The first of these is related to the limited participation of some of the focus group respondents, although that was constantly encouraged by the researcher who moderated these activities. The collection of data using the semi-structured interview technique could have revealed richer and more uninhibited perceptions on the part of these participants. On the other hand, the initial intention of the study to include undergraduate students ended in failure, as practically none of these agreed to participate (only one). It therefore seems pertinent for future research studies to focus on the comparison of the perceptions of both graduate and undergraduate students, from this faculty of Economics and Management, concerning the relevance of transferable skills and the contribution of the *Transferable Skills Development Programme,* since both groups benefit from this programme. Lastly, it would be useful to know the views of people who occupy leading positions in the companies from this region of Portugal and who often cooperate with the programme. These opinions would be useful in providing comments on the programme itself and, more especially, with regard to the transferable skills, they most value in their recently graduated collaborators.

References

1. Lawler, E. (1994). From job-based to competency-based organizations. *Journal of Organizational Behavior, 15*(1), 3–15.
2. Assiter, A. (1995). *Transferable skills in higher education*. Londres: Kogan Page.
3. Chadha, D. (2006). A curriculum model for transferable skills development. *Engineering Education, 1*(1), 19–24.
4. Drummond, I., Nixon, I., & Wiltshire, J. (1998). Personal transferable skills in higher education: The problems of implementing good practice. *Quality Assurance in Education, 6* (1), 19–27.
5. Mansfield, B. (2004). Competence in transition. *Journal of European Industrial Training, 28* (2/3/4), 296–309.
6. Raybould, J., & Sheedy, V. (2005). Are graduates equipped with the right skills in the employability stakes? *Industrial and Commercial Training, 37*(5), 259–263.
7. Stewart, J., & Knowles, V. (1999). The changing nature of graduate careers. *Career Development International, 4*(7), 370–383.

8. Binks, M. (1996). Enterprise in higher education and the graduate labour market. *Education + Training, 38*(2), 26–29.
9. Crammer, S. (2006). Enhancing graduate employability: Best intentions and mixed outcomes. *Studies in Higher Education, 31*(2), 169–184.
10. Fallows, S., & Steven, C. (2000). Building employability skills into the higher education curriculum: A university-wide initiative. *Education + Training, 42*(2), 75–83.
11. Gammie, B., Gammie, E., & Cargill, E. (2002). Personal skills development in the accounting curriculum. *Accounting Education: An International Journal, 11*(1), 63–78.
12. Harvey, L., Moon, S., Geall, V., & Bower, R. (1997). *Graduates' work: Organisation change and students' attributes.* Birmingham: Centre for Research into Quality together with the Association of Graduate Recruiters.
13. Nabi, G., & Bagley, D. (1999). Graduates' perceptions of transferable personal skills and future career preparation in the UK. *Education + Training, 41*(4), 184–193.
14. The European Higher Education Area. (1999). *The Bologna declaration of 19 June 1999: Joint declaration of the European Ministers of Education.* http://www.eurashe.eu/library/modernising-phe/Bologna_1999_Bologna-Declaration.pdf
15. Weil, S. (1999). Re-creating Universities for 'beyond the stable state': From 'Dearingesque' systematic control to post-dearing systematic learning and inquiry. *Systems Research and Behavioural Science, 16*(2), 171–190.
16. Sleezer, C. M., Gularte, M. A., Waldner, L., & Cook, J. (2004). Business and higher education partner to develop a high-skilled workforce: A case-study. *Performance Improvement Quarterly, 17*(2), 65–82.
17. Possa, G. (2006). Europe's Universities in response to Europe's challenges. *Higher Education in Europe, 31*(4), 355–357.
18. Andrews, J., & Higson, H. (2008). Graduate employability, 'soft skills' versus 'hard' business knowledge: A European study. *Higher Education in Europe, 33*(4), 411–422.
19. Garavan, T., & McGuire, D. (2001). Competencies and workplace learning: Some reflections on the rhetoric and the reality. *Journal of Workplace Learning, 13*(4), 114–164.
20. Sandberg, J. (2000). Understanding human competence at work: An interpretative approach. *Academy of Management Journal, 43*(1), 9–25.
21. Dench, S. (1997). Changing skill needs: What makes people employable? *Industrial and Commercial Training, 29*(6), 190–193.
22. McClelland, D. (1973). Testing for competence rather than for intelligence. *American Psychologist, 28*(1), 1–4.
23. Burgoyne, J. (1993). The competence movement: Issues, stakeholders and prospects. *Personnel Review, 22*(6), 6–13.
24. Moore, D. R., Cheng, M.-I., & Dainty, A. R. (2002). Competence, competency and competencies: Performance assessment in organisations. *Work Study, 51*(6), 314–319.
25. Woodruffe, C. (1993). What is meant by a competency? *Leadership & Organization Development Journal, 14*(1), 1, 29–36.
26. Strebler, M., Robinson, D., & Heron, P. (1997). *Getting the best out of your competencies.* Brighton: Institute of Employment Studies, University of Sussex.
27. Boyatzis, R. (1982). *The competent manager: A model for effective performance.* New York: Wiley.
28. Hoffmann, T. (1999). The meanings of competency. *Journal of European Industrial Training, 23*(6), 275–286.
29. Levy-Leboyer, C. (1996). *La Gestion des compétences.* Paris: Les Editions d'Organisation.
30. Spencer, L. M., Jr., & Spencer, S. M. (1993). *Competence at work.* New York: Wiley.
31. Lawrence, T. (2002). Teaching and assessing employability skills through skills USA. Quality Congress. ASQ's Annual Quality Congress Proceedings (pp. 285–295). Leesburg: Skills USA.
32. Tien, C., Ven, J., & Chou, S. (2004). Using problem-based learning to enhance students' key competencies. *Journal of American Academy of Business, 2*(2), 454–459.

33. Gibbons-Wood, D., & Lange, T. (2000). Developing core skills—Lessons from Germany and Sweden. *Education + Training, 42*(1), 24–32.
34. Atlay, M., & Harris, R. (2000). An institutional approach to developing students' "Transferable" skills. *Innovations in Education & Training International, 37*(1), 76–84.
35. Kearns, P. (2001). Generic skills for the new economy. *Review of Research.* Obtained in August 2015, from Review of research, NCVER. http://www.ncver.edu.au/publications/602.html
36. González, J., & Wagenaar, R. (2003). Tuning educational structures in Europe: Final Report Phase One. Obtained in August 2015, from the University of Deusto and University of Groningen. http://www.bolognakg.net/doc/Tuning_phase1_full_document.pdf
37. Billing, D. (2007). Teaching for transfer of core/key skills in higher education: Cognitive skills. *Higher Education, 53*(4), 483–516.
38. Gosling, J., & Mintzberg, H. (2006). Management education as if both matter. *Management Learning, 37*(4), 419–428.
39. Rubin, R., & Dierdorff, E. (2009). How relevant Is the MBA? Assessing the alignment of required curricula and required managerial competencies. *Academy of Management Learning and Education, 8*(2), 208–224.
40. Bennett, R. (2002). Employers' demands for personal transferable skills in graduates: A content analysis of 1000 job advertisements and an associated empirical study. *Journal of Vocational Education & Training, 54*(4), 457–476.
41. Farmer, J., & Campbell, F. (1997). Information professionals, CPD and transferable skills. *Library Management, 18*(3), 129–134.
42. Humphreys, P., Greenan, K., & Mcllveen, H. (1997). Developing work-based transferable skills in a university environment. *Journal of European Industrial Training, 21*(2), 63–69.
43. Laughton, D., & Montanheiro, L. (1996). Core skills in higher education: The student perspective. *Education + Training, 38*(4), 17–24.
44. Murray, S., & Robinson, H. (2001). Graduates into sales–employer, student and university perspectives. *Education + Training, 43*(3), 139–145.
45. Whittle, S., & Eaton, D. (2001). Attitudes towards transferable skills in medical undergraduates. *Medical Education, 35*(2), 148–153.
46. Rosenberg, S., Heimler, R., & Morote, E.-S. (2012). Basic employability skills: A triangular design approach. *Education + Training, 54*(1), 7–20.
47. Cabral-Cardoso, C., Estêvão, C., & Silva, P. (2006). *Competências transversais dos diplomados do Ensino Superior: Perspetiva dos Empregadores e Diplomados [Transferable skills of Higher Education graduates: Perspectives from Employers and Graduates].* Guimarães: TecMinho.
48. Hallinger, P., & Bridges, E. (2007). *Problem-based management education: Developing "managers for action".* Dordrecht: Springer.
49. Kitzinger, J. (1994). The methodology of Focus Groups: The importance of interaction between research participants. *Sociology of Health & Illness, 16*(1), 103–121.
50. Stokes, D., & Bergin, K. (2006). Methodology or "Methodolatry"? An evaluation of focus groups and depth interviews. *Qualitative Market Research: An International Journal, 9*(1), 26–37.
51. Powell, R. A., & Single, H. M. (1996). Focus groups. *International Journal for Quality in Health Care, 8*(5), 499–504.
52. Kitzinger, J. (1995). Qualitative research: Introducing focus groups. *British Medical Journal, 311*(7000), 299–302.
53. Grudens-Schuck, N., Allen, B. L., & Larson, K. (2004). Methodology brief: focus group fundamentals. *Extension Community and Economic Development Publications*, Book 12.
54. Jackson, D. (2009). An international profile of industry-relevant competencies and skill gaps in modern graduates. *International Journal of Management Education, 8*(3), 29–58.
55. Rainsbury, E., Hodges, D., Burchell, N., & Lay, M. (2002). Ranking Workplace Competencies: Student and Graduate Perceptions. *Asia-Pacific Journal of Cooperative Education, 3*(2), 8–18.

56. Robles, M. M. (2012). Executive perceptions of the top 10 soft skills needed in today's workplace. *Business Communication Quarterly, 75*(4), 453–465.
57. Quek, A. (2005). Learning for the workplace: A case study in graduate employees' generic competencies. *Journal of Workplace Learning, 17*(4), 231–242.
58. Keeling, R. (2006). The Bologna process and the Lisbon research Agenda: the European Commission's expanding role in higher education discourse. *European Journal of Education, 41*(2), 203–223.
59. Tomlinson, M. (2008). 'The degree is not enough': Students' perceptions of the role of higher education credentials for graduate work and employability. *British Journal of Sociology of Education, 29*(1), 49–61.
60. Cotton, K. (2001). *Developing employability skills*. Portland: Northwest Regional Educational Research Laboratory.
61. Hawkins, P., & Winter, J. (1995). *Skills for graduates in the 21st century*. Cambridge: Association of Graduate Recruiters.
62. Crebert, G., Bates, M., Bell, B., Patrick, C.-J., & Cragnolini, V. (2004). Developing generic skills at university, during work placement and in employment: Graduates' perceptions. *Higher Education Research & Development, 23*(2), 147–165.

Organizational Propensities to Share: Revisiting Talent Mobilization and Redistribution in Multinational Corporations

David Starr-Glass

Abstract This theoretical and reflective chapter examines human resource management practices regarding talent in multinational corporations. It explores the multiple meanings that have been attributed to talent and to its global management and then uses these meanings to question talent management practices within the organization. It suggests that current practices often reflect an implicit, but not fully articulated, set of philosophical assumptions (meta-principles) about talent and about the organization's propensity to internally share, mobilize, and redistribute its multiple talent resources. This chapter reviews ways through which appropriate talent can be mobilized within the multinational corporation and directed to areas where it can be optimally utilized. This chapter argues that relevant architectural structures and bridges should be created by the human resource function to facilitate the cross-border mobility and internal redistribution of organizational talent and considers the cross-cultural challenges and training involved.

1 Introduction

Talent has frequently been portrayed as a rare commodity that has to be aggressively battled for in the desolate wastelands of domestic and international marketplaces [1–4]. Perhaps, for some forms of talent and at some times in economic history, this has been true; however, the talent wars of the past tended to misrepresent the nature of talent as much as they overvalued the few and overpriced the efforts of the relentless talent warriors. It is perhaps more productive to view talent as something that is relatively abundant but which needs to be conscientiously searched for, thoughtfully nurtured, and strategically redeployed within the

D. Starr-Glass (✉)
University of New York in Prague, Prague, Czech Republic
e-mail: David.Starr-Glass@esc.edu

D. Starr-Glass
Empire State College, International Programs (Prague), State University of New York, Prague, Czech Republic

C. Machado (ed.), *Competencies and (Global) Talent Management*,
Management and Industrial Engineering, DOI 10.1007/978-3-319-53400-8_3

organization. But such an approach requires corporations, and global corporations in particular, to recognize talent as a living and vital asset that is dynamic and continuously developing—not as something that is exceptional in nature, acquired under difficult circumstances, and required only in exceptional times. As Mitchell [5] put it, 'global leaders don't appear spontaneously; the organization has a duty to create them' (p. 37).

Acquiring, nurturing, and developing appropriate talent are challenging propositions for any organization. For organizations with a presence beyond their national borders, the challenge is amplified because they operate in multiple contexts that are unfamiliar and which require a richer and more diverse range of talents and understandings [6, 7]. Fortunately, global organizations have a significant advantage because they can access multiple pools of human capital in the different geographic and cultural regions in which they operate. Still, global organizations, and more specifically multinational corporations (MNCs), have historically undervalued the human resources available to them in their 'peripheral' units, concentrating instead on the centralized development of talent in their headquarters and redistributing it through expatriation.

This chapter argues for a more dynamic approach to talent management in international and global organizational contexts. It suggests that talent can be effectively redistributed and mobilized through architectural structures and bridges that have been created by the human resource (HR) function of the multinational firm. Mobility possibilities, however, require an understanding that talent is an inherently fluid organizational resource that can be effectively circulated and repositioned within the global organization. Talent mobilization and redistribution, particularly in MNCs, might also need to be considered more broadly, not replying exclusively on the physical movement of organizational participants through expatriation and inpatriation [8, 9]. Talent mobilization can make a valuable contribution to solving problems in which change, strategic innovation, and talent development are complicated by national culture and unfamiliar marketplaces. But, for mobilization to materialize there needs to be a propensity to share talent and to utilize it fully within the organization [10, 11].

This chapter explores HR practice, arguing that implicit considerations and philosophies about talent are expressed in the MNC's talent management and that these philosophies provide some insight into an organizational propensity to mobilize and redistribute its talent resources. This chapter is structured in the following way. The first section explores the multiple meanings that have been attributed to talent and to its global management. The next section uses these meanings to reconsider the implicit assumptions and philosophies—what might be termed the *meta-principles*—which MNCs and their HR systems might hold about the fluidity and mobility of the talent that they possess. This is followed by a consideration of the architectural structures and bridges that the HR function might create to facilitate the cross-border mobility and utilization of organizational talent. The last section briefly summarizes the key issues presented in this chapter.

2 Organizational Talent and Its Global Management

At the outset, it might be instructive to consider the etymology of *talent*. In its original (Greek) sense, the word referred to a set of scales and by extension came to mean the process of weighing. Gradually, the word began to be connected with specific weights (particularly of silver) and in turn was used to denote money (i.e., coins of a designated weight). Later still 'talent' began to be used more generally as an attribution of value. The common thread in this progression of meanings is that neither weight nor value is absolute, inherent, or universally agreed upon. Both weight and value are defined relatively and determined contextually. Any weight can be designated as 'five units,' but only when that has been done will a weight that is twice as heavy be 'ten units'; the value of a glass of water, which is not the same as its price, varies according to the utility that it provides in a given context. The etymological origins of the *talent* are sometimes reflected in its contemporary use in the business and organizational worlds; more often than not, they are obscured. Talent is neither absolutely defined nor recognized; its value and relevance are contextual.

2.1 Toward a Definition of Talent

Reviewing the literature, some [12] have concluded that 'talent can mean whatever a business leader or writer wants it to mean, since everyone has his or her own idea of what the construct does and does not encompass' (p. 291). It is certainly true that talent has been defined in multiple ways, with definitions reflecting specific aspects that individual researchers and practitioners have considered uniquely important. More significantly, these definitions include a range of different philosophical assumptions about the nature of talent, even though many would prefer that these definitions were more firmly embedded in a matrix of objective psychology [13].

For example, Tansley et al. [14] have suggested that talent requires a consideration of *individual giftedness* and 'can be considered as a complex amalgam of employees' skills, knowledge, cognitive ability, and potential' (p. 2). Others, such as Bethke-Langenegger [15], understand talent not as an abstraction or potential, but rather as an *embodiment*: 'one of those worker who ensures the competitiveness and future of a company (as specialist or leader) through his [sic] organisational/job specific qualification and knowledge, his social and methodical competencies, and his characteristic attributes such as eager to learn or achievement oriented' (p. 3). Yet others, such as Ulrich [16], view talent as not simply the embodiment of a set of personal attributes, but rather as a triple expression of these attributes within the organizational context—individual *competencies*, coupled with a future-orientated organizational *commitment*, and providing a significant organizational *contribution*.

Although not wishing to contribute to the plethora of definitions, in this chapter talent is understood as that set of personal and relational attributes, qualities, and capacities—possessed by an individual, and recognized and assessed by the organization—that are considered of value in contributing to the organization's operational and strategic goals. This definition suggests that talent:

- is a complex set of personal attributes, skills, and competencies that are presently recognizable;
- also includes capacities and potentials that can probably be further developed;
- includes social and relational abilities as well as firm-specific operational skills;
- has been identified by the organization, not simply by the individual or by others outside the organization; and
- is considered of sufficient value and importance to be recruited, selected, developed, retained, and utilized by the organization in realizing its current and future goals.

2.2 Talent Management and Global Talent Management

The term *talent management* originated from the practitioner literature and many researchers and scholars have argued that the term is imprecise and lacks an accepted definition; indeed, some suggest that the subject is a minefield of 'rhetorical obfuscation' [17–21]. Beginning their review of the construct, Cappelli and Keller [18] note that 'nearly every article written on the topic begins with hand-wringing over the conceptual boundaries of the term' (p. 306). In keeping with that tradition, it might be appropriate at this point to indulge in some hand-wringing, albeit in a restrained and limited way.

For Reilly [22], it appears that *talent management* has always been a tricky subject, 'at risk of becoming mere hyperbole… or of becoming the fad of the conference circuit because the term lacks a clear definition… definitions are, at worst, a mélange of different concepts strung together without a clear statement of what is meant by talent and how we might manage it' (p. 381). Collings et al. [23] are also of the opinion that 'the concept of talent management is lacking in terms of definition and theoretical development and there is a comparative lack of empirical evidence on the topic' (p. 1264). Of course, lack of a clear and convincing definition might simply reflect the growing pains of talent management as it moves from its infancy to adolescence [24].

A recent comprehensive review by Cappelli and Keller [18] proposes a definition of talent management which they claim 'is consistent with traditional approaches and captures what academic researchers have been doing under the heading of talent management… the process through which organizations anticipate and meet their needs for talent in strategic jobs' (p. 307). This is a broad and somewhat bland definition, but it does accentuate the anticipated strategic-related outcomes of the talent management process, even though it provides little guidance as to the nature of the process itself.

Talent management is usually, but not exclusively, considered to be an aspect of the organization's ongoing HR activity. Optimally, considerations of talent management should permeate all aspects of the organization—particularly its senior-level management and governance structure—even though talent management might be operationally centered in the HR function. Talent management can be particularly challenging when the organization operates across national and cultural borders. Operating across such borders is explicit (a) for global or multinational corporations, which operate in a number of countries but have a designated home country in which they are headquartered and from which they expand, and (b) to a somewhat lesser extent, for transnational corporations that have operations more evenly dispersed throughout multiple countries. In these border-crossing contexts, talent management has to deal with issues of global expansiveness, both in terms of accessing talent-related opportunities throughout the firm's entire operational range and in recognizing and developing talent that would not have been available domestically [25].

Like talent and talent management, *global talent management* (GTM) has also been variously recognized and defined [20, 23, 26, 27]. Some scholars and practitioners regard GTM in fairly general terms, extending 'domestic' talent management concepts into the global arena. Caligiuri et al. [28], for example, see GTM as including all of those generally recognized HR activities—attracting, recruiting, selecting, training, developing, and retaining—focused on the 'best' employees, in order to achieve organizational strategic priorities on a *global scale*, taking into account the differences in the organization's global strategic priorities and in the national contexts within which the organization operates.

Other definitions of GTM have emphasized the dynamic and constantly *changing context* within which it is practiced. Thus, Tarique and Schuler [20] see GTM as the systematic utilization of 'international human resource management' in order 'to attract, develop, and retain individuals with high levels of human capital (e.g., competency, personality, motivation) consistent with the strategic directions of the multinational enterprise in a dynamic, highly competitive, and global environment' (p. 124). This definition serves as a useful description of what global talent management is—and of what it might be—but the pressing question is how such a definition might inform the talent management process within organizations that are international in nature, or which purport to be global in their reach. Put differently, how might MNCs recognize and align their talent assumptions, philosophies, and practices in ways that lead to effective GTM?

3 Interrogating Organizational Meta-principles Regarding Talent

MNCs have an active presence in a number of different countries; as such, they confront two challenges. First, each regional center must operate effectively within its immediate task environment, recognize local opportunities and threats, and respond and adapt to country-specific conditions. Second, the MNC must also preserve a sense of organizational continuity in terms of a consistent culture, coordinated strategic direction, and integrated leadership and governance throughout all of its differently located units. These two forces—one centrifugal, diverging, and moving toward fragmentary difference; the other centripetal, converging, and moving toward unifying similarity—must be continuously balanced. As part of that balancing process, each MNC must uniquely decide on the degree to which its headquarters and national units are centralized, coordinated, or autonomous in their operations [29].

So far as talent is concerned, the MNC presents an interesting case. The international reach of their different national presences provides the MNC with the opportunity to recruit, develop, and retain talent that would not have otherwise been available domestically. However, two understandings are required to capitalize on this opportunity. First, talent has to be seen as spread widely and generally, not as being limited, restricted, or aggregated in specific geographic regions—particularly the organization's headquarters and its home country. Second, talent needs to be understood not as an abstract personal attribute or potential capacity, but as something that is contextually determined, situationally salient, and differently recognized [5, 24].

It is suggested that a deeper exploration of these competing assumptions about talent—assumptions that are sometimes not clearly articulated but which are implicitly embedded in talent management practice—can provide the MNC with a better understanding of how it *really understands* talent, how it might effectively mobilize and deploy its global talent, and how it might be possible to create a comprehensive HR architecture within the organization that can facilitate such redistributions [30, 31].

3.1 Inclusive and Exclusive Talent Perspectives

Talent management can be approached from either an inclusive or exclusive perspective. These perspectives reflect quite differing understandings of what constitutes talent. But more than that, they also communicate different understandings of what constitutes organizational culture and of what organizational membership, organizational participation, and organizational contribution might mean. At the outset, it is important to bear in mind that although inclusive and exclusive approaches have usually been considered binary opposites—with exclusive versions being generally more favored—there is an increased understanding that they are not mutually contradictory and might be more productively viewed as polar positions along a spectrum of possibilities [32–34].

Inclusive Talent Perspectives From these perspectives, talent is recognized as existing widely within the organization and all organizational participants are included in the process of talent management [35]. Thus, inclusive approaches provide all organizational participants with an *equal opportunity* to demonstrate talent, even though some may later fail to do so and be subsequently removed from the talent pool. As Swailes and Downs [36] note, fully inclusive talent management 'is the recognition that all employees have talent, together with the ongoing evaluation and deployment of employees in positions that give the best fit and opportunity for employees to use those talents' (p. 6). They add that talent should be evaluated through a process of consultation between employees and their organization. When talent levels fall below the standards that have been set, there is an expectation that the organization will assist employees to develop their talents, use them in more effective ways, or deploy them in different situations in which their talents might be better optimized. In inclusive talent management approaches, there is also an understanding that the standards for evaluating talents will be set 'democratically, not arbitrarily.'

Crucially, inclusive talent perspectives recognize that all organizational participants have been recruited, selected, and hired *because* they possess talents that are valuable for the enterprise. In that sense, inclusiveness is as much about the recognition of legitimate organizational belonging and contributing as it is about the actual recognition of talent. The way in which the organization recognizes and appreciates talent is reflected in the organization's utilization of these resources for its present and future growth. However, the organization also remains sensitive to how those possessing talent view themselves, to their sense of contributing to the workforce, and to their perception of organizational belonging, trust, and loyalty. As Downs and Swailes [37] put it: 'Talent identification should encourage people to consider and to realize what matters to them in line with the interests of the organization' (p. 277).

Exclusive Talent Perspectives Historically, the dominant approach has been an exclusive perspective in which: (a) talent is understood to reside with the few in the senior levels of the organizational hierarchy; and (b) talent management is devoted to the acquisition, retention, and promotion of those identified few [38]. Exclusive approaches are often linked to contexts in which there is high workforce differentiation, and where employees perform distinct and segregated activities [39, 40]. Exclusive approaches also understand that talent is a scare resource and that its distribution is highly skewed—certainly in the general population, and most likely in organization. With these assumptions in mind, it has seemed more efficient, in terms of organizational resources, to focus only on those who have been identified as possessing high talent potential. It is always difficult to calculate rates of return on talent, but some have questioned the putative higher rates of returns generated by investing exclusively on those that have those identified as having high talent potentials [41].

3.2 Internal and External Sources of Talent

Related to inclusive and exclusive talent perspectives is the relative importance that the organization places on acquiring new talent from internal or external sources. Sometimes, a review of existing talent within the MNC will reveal that there is no good fit between what is available and what is required; in such cases, external hiring is obviously necessary. Sometimes, however, an internal search will suggest that suitable talent is in fact presently available within the organization and that promotion, transfer, or some other forms of internal redistributions might be appropriate. Clearly, there is no single perspective that is optimal; most organizations will utilize both, albeit intuitively preferring and stressing one. External hiring or internal mobility decisions are not always the binary choices dictated by logic: They represent deeply held perspectives that the organization and its HR function hold about talent, how it is developed, and how it is retained.

Internal Sources of Talent Looking toward external sources may be the only reasonable alternative for organizations that do not possess—or do not believe that they possess—the talent they desire. External sources might provide a solution, but the decision to recruit new talent from outside the organization begs the question of *why* that talent is not available internally. The answer might be straightforward and reasonable: a novel situation, an unprecedented talent requirement, or a similar unexpected event. However, the reflexive reliance on external talent can indicate an implicit HR philosophy of how talent is developed, cultivated, matured, and mobilized within the organization.

Historically, organizations have tended to underappreciate their internal talent sources. More recently, there has been growing interest in the possibilities of internal cultivated and redistributed talent [42]. Increasingly, MNCs are recognizing that talent and innovation are not necessarily location-specific and that they can draw upon the diversified portfolio of talent that they manage [43]. As Cantwell and Zhang [44] note, because talent is 'in part location-specific as well as firm-specific, the MNC has come increasingly to draw upon a diversified locational portfolio of capabilities' (p. 46).

This requires MNCs to recognize the talent diversity and richness that they can presently access; still, talent diversity per se is not necessarily an advantage. The organization has to ensure that it has *already* selected, recognized, and cultivated the appropriate talent that it needs, or believes it will need, within these pools of diversity. Further, once it is recognized that talent is not location-specific, the MNC needs to create ways of mobilizing and directing it toward the purposes and places where it might be optimally utilized [45, 46].

External Sources of Talent The hiring of non-organizational candidates provides an injection of talent that is believed to be lacking in the organization. It gives the organization an opportunity to appropriate talent that it considers valuable, which has been developed in other organizations, and which is now available in the external labor markets [47, 48]. But, in times of uncertainty, external hiring is associated with a number of inherent risks: asymmetric information, the subsequent suitability and long-term performance of the individual, and his/her possible failure to adapt successfully to the new firm [49]. Nevertheless, many organizations—depending on the industrial sector and the nature of the position—find that external sources of talent offer flexible, convenient, 'just-in-time' solutions for their needs. Organizations may also search for external talent without the intent of permanently incorporating it in their structures. The growing externalization of labor and the changing nature of work and employment—outsourcing, co-employment, freelancing, and independent contracting—have significantly challenged the traditional assumptions about the 'ownership' of talent and the ways in which it is utilized [50–52].

3.3 *Talent of People or Talent for Jobs*

When faced with acquiring talent, organizations can proceed in two different ways. As Cappelli and Keller [18] put it: 'Should we begin by identifying a subset of individuals who might be slotted in an array of roles, or by identifying a specific population of jobs that are in some way strategic and then focus on filling those roles with talent?' (p. 306). Either of these strategies might be appropriate, depending on the circumstances and the nature of the organization. Firms should be flexible and pragmatic in deciding which approach to adopt; nevertheless, approaching talent from either a job or person perspective may signal predispositions in the ways that the organization conceptualizes talent.

Talent of People Lepak and Snell [53] suggest that 'the value of human capital is inherently dependent upon its potential to contribute to the competitive advantage or core competence of the firm' (p. 35). The people in an organization constitute a critical resource but, unlike financial or material resources, human capital is rented temporarily and not permanently owned by the firm. Focusing on people in the recruitment and selection process provides a flexible source of high-quality human capital for the organization that can potentially increases its competitive advantage and core competence. Focusing on people in the development and utilization process offers the organization the possibility of retaining and having ongoing access to the human capital that it has acquired [40]. As Becker and Huselid [54] note, there is a difference between the present and future value of employee talent. Present value might, they suggest, be understood as a supply-side phenomenon, whereas future value 'is a function of how those skills are used and where they are used' (p. 904).

Talent for Jobs Job-led recruitment and selection involves defining the hole and then looking for the most appropriate peg that fits it. If the 'hole-like' quality of the job is considered appropriate, the traditional approach has been to analyze the job and then suggest the kinds of knowledge, skill, ability, and other characteristics that candidates would need to successfully fit the defined that job or the position [55, 56]. The problem is that jobs are not static holes, any more than and people are wooden pegs. The job-centered approach makes the vital talent possibilities of individuals curiously subservient to the talent requirements of inanimate jobs. It also accentuates the presumed stability of the job.

Although job analysis—which privileges the importance and permanence of the job itself—persists, it has gradually been overtaken by a more dynamic appreciation of the importance of the employee and of his/her *understanding and management* of the job. As Sanchez and Levine [57] observe 'the scope of job analysis research is being expanded toward a better understanding of work demands as experienced by job incumbents, both individually and collectively through shared perceptions' (p. 417).

Situations will always exist in which organizations have to fill specific firm-related jobs and will recruit and select candidates with the defined position foremost in mind. This can be a rational and pragmatic expediency, but filling the job is not the same as expanding or enriching the organizational talent pool. Developing and nurturing the people who form the talent pool reduces the organization's concern about future talent needs and optimize its operational flexibility.

3.4 Reconsidering Underlying Human Resource Philosophies

Logically, it can be argued that practices emerge from underlying assumptions and philosophical considerations: *What* is done and *how* it is done are rooted in *why* it is being done. However, HR practices often surface as the piecemeal ad hoc solutions for discrete and unrelated problems, resulting in an uneven mosaic that covers—rather than reveals—the fundamental philosophies related to people, talent, and work.

Schuler [58] defines HR philosophies as the broad and organizationally accepted understandings of how 'how the organization regards its human resources, what role the resources play in the overall success of the business, and how they are to be treated and managed' (p. 21). Yet, organizations can fail to reflect on their deep-seated assumptions about the human resources available to them or to appreciate the role that their *meta-principles* of talent play in current and future operations. Developing and interrogating the philosophical assumptions of HR is not an abstraction; it is a crucial element in sustaining competitive advantage and future success. As such, HR assumptions need to be constantly and critically appraised [59–61].

For any organization, it is unproductive and frustrating to believe that talent exists as a hidden potential in unknown places. For the MNC, there is the possibility that talent can be 'discovered' and utilized throughout the global range of its operational presence. But to recognize and deploy organizational talent effectively global firms, their senior management, and their HR professionals must question a number of related assumptions, philosophical perspectives, and meta-principles:

- Do we tend to recognize talent inclusively or exclusively?
- Do we presume that talent is more likely located internally or externally? and
- Do we usually consider talent to be an attribute of individuals or to be connected with specific jobs and positions?

On reflection, it might be that organizations find that they place higher value on inclusive, internal, and people-centered understandings about talent, even though their HR practices do not fully or consistently reflect these values. It is suggested that valuing inclusive, internal, and people-centered understandings about talent demonstrates that the organization has a higher *propensity* to consider the redistribution and mobilization of talent resources. This does not necessarily mean that exclusive, external, and position-centered assumptions will *not* lead to talent mobility. However, where exclusive, external, and position-centered assumptions about talent prevail, it is suggested that talent mobilization—particularly in MNCs—might be limited and restricted to specific (and costly) expatriations. It is also likely that organizational structure—particularly the perceived dominance of the MNC headquarters—will significantly moderate the organization's propensity to mobilize and internally redistribute talent.

It seems valuable for the effective operation of the MNC and for the success for its strategic development that: (a) there is a critical evaluation of its current HR appreciations of the nature and possibilities associated with the organizational talent; and (b) the revealed philosophical assumptions and meta-principles are then consistently and transparently expressed in the HR policies and processes of the organization. This requires a continuous effort to align what is intended with what is done. Although the primacy of HR is frequently assumed—particularly by those in the HR function—it is all too easy for significant gaps to exist between the *intended* processes that the HR function espouses, the *enacted* processes that are brought into existence, and the *perceived* processes that organizational participants experience [62].

There may be a plurality of approaches to HR and to talent management within the organization; however, the value of the HR function cannot become evident unless it has thoughtful constructed architectural structures and linkages between organizational culture, perceived organizational climate, and the ways in which people and their talents are managed. HR must convincingly articulate its guiding philosophies, develop appropriate policies, sustain effective practices, and implement meaningful processes that provide advantages for the organization and its participating members [63, 64]. In MNCs specifically, there is a clear need for a thoughtful and well-designed HR architecture that connects and integrates the geographically (and sometimes culturally) separated parts of the organization.

4 Architectural Structures for Talent Mobility and Utilization

The HR function may be centrally located, but to operate effectively the management of talent needs to be comprehensively directed across the global organization with an adequate consideration of the specific national contexts and situations in which talent is located. If talent is to have any significance or value, GTM needs to explore and facilitate the optimization of mobility within the organization's disparate talent pools. This requires the HR function to creatively develop architectural structures and bridges to ensure that appropriate talent is recognized and nurtured locally and is then made available wherever it is needed in order to meet the operational and strategic requirements of the whole organization [30, 54].

For the MNC, the ability to access the internal talent that is within its operational reach is particularly important, because the search for new talent—talent that is overlooked or neglected—is one of the significant drivers in the geographic relocation of core parts of the corporation, relocations that can be costly and entail a considerable amount of risk [65, 66].

4.1 Expatriation: Connecting Core and Periphery

Talent mobility seems a logical response for the MNC; however, the nature and purpose of mobility is mediated by the prevalent HR philosophies about talent and its management. Cerdin and Brewster [67] observe that there are two streams of thought as follows: '(1) the elitist, or talent segmentation, approach focused on a few chosen individuals, often termed "high potentials", in whom the organization invests, and (2) a broad aspect of Human Resource Management where all employees are considered as talent' (p. 245). Following the first stream of thought, intrafirm mobility via expatriation—that is moving employees from the MNC's headquarters, or core, outwards to 'peripheral' units—is seen as providing a few high-value individuals with experiences that will be beneficial for the long-term strategic goals of the organization.

Under the second broader understanding of *inclusive talent*, expatriation is regarded [67] 'as an invaluable developmental experience offered to employees being assigned abroad, rather than restricted to just the declared "developmental" assignments, and is seen as an additional "weapon" in the "armory" of talent-management specialists' (p. 247). Using this approach, mobility seeks to promote a more diffuse experiential richness for individuals and for the organization as a whole, rather than to limit it to those that have been identified as having high talent, potential, or power. Further, initiating and promoting the transfer of MNC participants demonstrates an organizational desire to integrate HR practice and to further internationalize the existing organization [68, 69].

Reviewing the global mobility literature, Caligiuri and Bonache [70] identify three main reasons for expatriation prior to the 1990s: (a) exercising of control over units of the MNC; (b) filling existing positions; and (c) providing management development. Since the late 1990s, organizations have increasingly been confronted with the reluctance of organizational participants to commit to long-term stays abroad, the recognition that expatriates may experience transcultural difficulties, and the growing appreciation of the significant expenses, direct and indirect, of traditional expatriate arrangements. All of these factors have influenced the traditional concept of the expatriation and have forced a re-evaluation of the benefits and value added to the organization by this kind of mobility [71].

In the last twenty-five years, there has been a growing trend for expatriate placement to be of shorter durations, revolving around a specific project, and with the expatriate not to be accompanied by family or significant others [72–75]. In economic environments that have become more challenging and competitive, employees have become more aware of their career vulnerability, of their own mobility, and of the developmental advantages that mobility offers. This has resulted in a greater employee interest in accepting short-term assignments abroad, leading PricewaterhouseCooper [76] to suggest that firms and their participants should view talent redistribution as 'a more fluid concept, driven by strategic need but also by the desire to optimise the investment in a mobility programme' (p. 11).

4.2 *Inpatriation: Connecting Periphery and Core*

It might be worthwhile to point out that 'expatriation' literally means to take individuals out from what they regard as their home (Latin, *pater* = father) country. Similarly, 'inpatriation' moves the person in to what will *become* his or her new 'fatherland.' Although these terms are commonly used in the GTM literature, they raise an interesting question for MNCs: To what extent does the MNC believe that there is a single dominant country in their global operations? Put another way, does the organization see itself as essentially a unified and holistic entity or as a federation of discrete and different units with the organization's headquarters being dominant?

Intrafirm talent flow from what might be regarded as the 'periphery' of the MNC toward the organizational center is a relatively recent phenomenon [77, 78]. As has been noted, the older notion of outward-bound expatriation placed significance on the exercise of control of the organizational headquarters over its more distant units. Inward inpatriation does not revolve around issues of control; instead, it revolves around issues of organizational inclusiveness and diversity.

Inpatriation recognizes that inbound organizational members possess different knowledge and understanding, and that their talents that can contribute to the corporation's core. Reiche [8], for example, notes that 'inpatriates can be more effective knowledge conduits by highlighting the importance of their cross-unit boundary spanning for knowledge transfer' (p. 383). Further, as Reiche et al. [79] argue since inpatriates have an 'intimate understanding of both the HQ and the local subsidiary context … they are able to cross existing intra-organizational, cultural, and communications boundaries to diffuse information' (p. 160).

Inpatriation and expatriation are in many ways symmetrical, with inpatriation reflecting all of the challenges associated with expatriation. The incomer—similar to the expatriate—may well experience language and cultural difficulties, be treated as a member of an out-group, disadvantaged in demonstrating the talent potential that he/she truly possesses, and lack the interpersonal trust of those that he/she works with [80, 81]. Just as with expatriates, incoming organizational participants may also have to deal with the difficulties of acculturation and stress in the new position, and with adapting to a 'new' corporation, even though it ostensibly has the same organizational culture as the unit from which they have come [82, 83].

Inpatriation and expatriation provide significant flows of talent throughout the MNC. They present opportunities for the redistribution and reassignment of talent. They bring relevant skills, competencies, and experience to places in the organization where they can be most effectively utilized. Talent mobility is also connected to new learning possibilities, allowing for the transfer of new knowledge and understandings when employees have completed their assignments.

HR initiated transfers require considerable resources and support for the transferred individual. However, they also provide a sense of continuity, community, and extensiveness within the organization—outcomes that have organizational value, even though the extent of that value might be difficult to quantify. Although there are probably more substantial benefits to be derived from long-term transitions, short-term stays and rotations can generate significant organization and personal value [84, 85].

4.3 Global Talent Teams: Virtual Redistribution

Traditionally, talent mobility has been connected with the physical movement of the person possessing that talent. With the extraordinary development of communication technologies and computer-mediated communication (CMC), however, talent can now be utilized and redistributed electronically and virtually.

By definition, MNCs operate in and are confronted with *discontinuities*. Discontinuities are factors such as physical locations, time zone differences, and differing national cultures that fracture continuity and tend to reduce organizational cohesiveness [86–88]. Discontinuities can potentially localize the otherwise extended operations of the organization, disrupt flows of information, and inhibit the mobility of talent. Since the 1990s, it has become increasingly common for organizations to use CMC to bridge the natural discontinuity that they encounter, recognizing that the distributed efforts of separated contributors can be effectively brought together into a collective whole. Bringing distributed talent together via virtual work group solves many logistical issues, but it also presents difficulties. The dynamics of managing communication and the processes of sharing knowledge differ from face-to face meeting partly because of channel richness, but even if channel richness is optimal, participants still have to deal with different national cultures and with the 'common' working language used by the work group [89, 90].

National culture difference can become apparent in the different assumption, beliefs, and accepted behaviors that group members hold. For instance, national culture differences can exist in dimensions such as power differences, notions of individualism and collectivism, the avoidance or acceptance of uncertainty, and differences in long or short-term orientations [91]. All of these nationally held assumptions, values, and norms can shape and potentially distort the sharing of knowledge [92, 93]. Individual national culture differences between team members and the national composition of the team also impact its outcomes, but as yet there is no clear understanding as to how these elements interact or moderate one another [94, 95]. Of particular interest are the different ways in which national culture might moderate different understandings of *virtuality* and of the process of collaboration in virtual environments [94, 96].

Additional problems and challenges can present themselves in the language used by the virtual group. Although English tends to be the most commonly used language, participants may possess significantly different levels of language understanding and comprehension that may adversely impact the group's performance and knowledge sharing effectiveness. Virtual environments—even those that have significant channel richness and which incorporate synchronous visual elements—tend to reduce many of the nuances of communication and intended meaning [97–99].

Global work teams provide a simple but effective way of mobilizing talent within the MNC. Their virtual nature means that the participants only have a partial immersion in the stimulating difference that exists within the different parts of the organization. Nevertheless, the virtual redistribution of talent can provide significant advantages for participants and the MNC [100, 101]. Additionally, the implementation of virtual global networks offers a significant HR architecture for facilitating talent mobility and signals, an organizational commitment to recognizing, rewarding, and nurturing that talent which it possesses.

4.4 Cross-cultural Competency Training

Because of the global operations of the MNC, talent mobilization almost inevitably means crossing of national culture boundaries, if not actual-state borders. The redistribution of talent across cultural boundaries poses challenges, and the MNC must ensure that relocated or redistributed organizational participants receive adequate training to function effectively.

Training has often focused exclusively on expatriates. In part, this can be explained by the expense associated with the long-term expatriation and the need to provide cross-cultural training to ensure that this investment succeeds. But the focus on expatriate training has also reinforced the notion that the MNC headquarters is central, dominant, and possesses cultural preeminence. Outbound migrations of organizational participants were thus seen as ventures into a peripheral zone that was less understood and that such ventures needed considerable organizational support [102–104]. However, in MNCs, *all* mobilization and internal redistributions of talent—whether in the form of expatriation, inpatriation, or virtual networks and team initiatives—involve movements across national cultures, and appropriate cross-cultural training is important for all who participate in these experiences.

In developing its cross-culture training initiatives, the HR function needs to critically identify the 'cultures' that redistributed organizational participants will actually negotiate. To some extent, all organizational participants share a common organizational culture, which Schneider et al. [105] define as a set of 'shared basic assumptions, values, and beliefs that characterize a setting and are taught to newcomers as the proper way to think and feel' (p. 362). However, organizational members view this culture through the prism of different national cultures, which may introduce distinct colors, refractions, and distortions of what is assumed to be a 'common' and unifying culture. The challenge for the HR function of the MNC is: (a) to identify the importance and impact of the distinctive national culture dimensions that might lead to confusion or misunderstanding within the organization; and then (b) to provide organizational members with the understanding, skills, and competencies required to negotiate the foreign cultures that will surround them when they are relocated or work with others.

In developing talent that can be effectively shared and mobilized within the MNC, the HR function sometimes places undue emphasis on the presumed permanence and static nature of national cultures that are to be negotiated. It is important to re-evaluate this, recognizing existing measures of national culture difference are statistically derived and there is frequently more in-country variation in cultural values than between-country variation [91]. Indeed, expatriate training can result in 'sophisticated stereotyping' rather than the ability to successfully negotiate, communicate, and actually enjoy the culture of the target country [106]. In approaching the challenge of talent nurturing, development, and mobilization, the HR function might find it more effective to reconsider the dominance of static approaches to national culture and to begin exploring the diversity and fluidity of culture boundaries that expatriates, inpatriates, and virtual team members actually encounter [94, 107, 108]. Similarly, HR might want to monitor the actual experiences and challenges of those who have participated in mobilization and continuously revise training initiatives to address problems and concerns that have been identified.

5 Conclusion

Organizations are continuously challenged to come to an understanding of what constitutes talent. Their definitions might shift over time, and they may embrace an eclectic set of understandings that change and assume a new holistic meaning over time. However, all organizations are crucially dependent on the talent resources that they possess and to better appreciate those resources and deploy them effectively there needs to be a clear operational understanding of what talent is and about how it can be acquired, developed, retained, and utilized. Some have struggled with the diverse meanings that have been attributed to talent, fearing that it might actually defy definition [12]. But organizations and their HR functions have to move beyond the philosophy of speculation and determine their own firm-specific talent meanings and then explore the consequences of those meanings.

Given the geographic and cultural reach of the transnational firm and the MNC, it seems likely that they have the possibility of finding and developing talent that is not available to the domestic firm. But having access to talent is not the same as recognizing it; having potential opportunities is not the same as realizing them.

It may well be that an inclusive talent reorientation would allow MNCs to recognize new and unique talent and also to acquire it. It has been suggested that the internal talent perspective is essentially a *chimera*—and intriguing, fantastical, but insubstantial mirage—because talent cannot reside in everyone. The argument set out in this chapter, however, is that talent *does* reside in all organizational participates *when* they have been purposefully recruited, nurtured, and retained because of their talent contribution.

Talent should not be thought of as an afterthought; something that is frantically looked for only occasionally when it is recognized that it is needed. Talent considerations need to be forefront in all organizational recruitment and selection decision—part of a continuous, forward-looking, process. As Swailes and Downs [36] have observed, 'talent has to be treated as an absolute characteristic of individuals and as something that needs to be identified and deployed in roles that map onto present and potential talent' (p. 13).

If an inclusion-centered appreciation of talent emerges as the organization's fundamental meta-principle concerning human capital and resources, it is more likely that a propensity for the intrafirm mobilization of talent will also materialize. Intrafirm talent mobilization offers the organization the possibility of more effectively deploying the talent richness that it possesses. Transnationals and MNCs have exceptional opportunities for amassing particularly rich pools of talent in the different national and cultural areas within which they operate. This richness of talent diversity may not be recognized or appreciated. However, even if recognized and appreciated, talent cannot be operationally or strategically deployed unless there is a robust HR architecture that facilitates the mobility, redistribution, and reassignment of that talent. It is to be hoped that more MNCs realistically reappraise their propensity to share talent more widely throughout the full geographic range of their operations, and revisit the opportunities—often unseen and often lost—of talent mobilization and redistribution [59, 109, 110].

References

1. Beechler, S., & Woodward, I. C. (2009). The global 'war for talent'. *Journal of International Management, 15*(3), 273–285.
2. Chambers, E. G., Foulon, M., Handfield-Jones, H., Hankin, S. M., & Michaels, E. G., III. (1998). The war for talent. *McKinsey Quarterly, 1*(3), 44–57.
3. Gardner, T. M. (2003). In the trenches at the talent wars: Competitive interaction for scarce human resources. *Human Resource Management, 41*(2), 225–237.
4. Somaya, D., & Williamson, I. O. (2011). Embracing turnover: Moving beyond the 'war for talent'. In H. Scullion & D. G. Collings (Eds.), *Global talent management* (pp. 74–86). Abingdon, UK: Routledge.
5. Mitchell, S. (2013). *Talent beyond borders: An organizational guide to delivering the promise of global talent management.* Pittsburg, PA: Development Dimensions

International. Retrieved from http://www.ddiworld.com/DDI/media/trend-research/talentbeyondborders_tr_ddi.pdf?ext=.pdf
6. Javidan, M., Teagarden, M., & Bowen, D. (2010). Managing yourself: Making it overseas. *Harvard Business Review, 88*(4), 109–113.
7. Mendenhall, M., Reiche, B. S., Bird, A., & Osland, J. S. (2012). Defining the 'global' in global leadership. *Journal of World Business, 47*(4), 493–503.
8. Reiche, B. S. (2011). Knowledge transfer in multinationals: The role of inpatriates' boundary spanning. *Human Resource Management, 50*(3), 365–389.
9. Reiche, B. S. (2012). Knowledge benefits of social capital upon repatriation: A longitudinal study of international assignees. *Journal of Management Studies, 49*(6), 1052–1077.
10. Schuler, R. S., Jackson, S. E., & Tarique, I. (2011). Global talent management and global talent challenges: Strategic opportunities for IHRM. *Journal of World Business, 46*(4), 506–516.
11. Preecea, D., Ilesb, P., & Jones, R. (2013). MNE regional head offices and their affiliates: Talent management practices and challenges in the Asia Pacific. *The International Journal of Human Resource Management, 24*(18), 3457–3477.
12. Gallardo-Gallardo, E., Dries, N., & González-Cruz, T. F. (2013). What is the meaning of 'talent' in the world of work? *Human Resource Management Review, 23*(4), 290–300.
13. Nijs, S., Gallardo-Gallardo, E., Dries, N., & Sels, L. (2014). A multidisciplinary review into the definition, operationalization, and measurement of talent. *Journal of World Business, 49*(2), 180–191.
14. Tansley, C., Harris, L., Stewart, J., & Turner, P. (2006). *Talent management: Understanding the dimensions. Change Agenda*. London, UK: Chartered Institute of Personnel and Development.
15. Bethke-Langenegger, P. (2012). *The differentiated workforce: Effects of categorization in talent management on workforce level.* Unpublished working paper, No. 18, Department of Business Administration, University of Zurich, Switzerland.
16. Ulrich, D. (2007). The talent trifecta. *Workforce Management*, September 10, pp. 32–33. Retrieved from http://www.workforce.com/articles/the-talent-trifecta
17. Huang, J., & Tansley, C. (2012). Sneaking through the minefield of talent management: The notion of rhetorical obfuscation. *The International Journal of Human Resource Management, 23*(17), 3673–3691.
18. Cappelli, P., & Keller, J. R. (2014). Talent management: Conceptual approaches and practical challenges. *Annual Review of Organizational Psychology and Organizational Behavior, 1,* 305–331.
19. Gallardo-Gallardo, E., Nijs, S., Dries, N., & Gallo, P. (2015). Towards an understanding of talent management as a phenomenon-driven field using bibliometric and content analysis. *Human Resource Management Review, 25*(3), 264–279.
20. Tarique, I., & Schuler, R. S. (2010). Global talent management: Literature review, integrative framework, and suggestions for further research. *Journal of World Business, 45*(2), 122–133.
21. Dries, N. (2013). Talent management, from phenomenon to theory: Introduction to the Special Issue. *Human Resource Management Review*, *23*(4), 267–271.
22. Reilly, P. (2008). Identifying the right course for talent management. *Public Personnel Management, 37*(4), 381–388.
23. Collings, D. G., Scullion, H., & Dowling, P. J. (2009). Global staffing: A review and thematic research agenda. *International Journal of Human Resource Management, 20*(6), 1253–1272.
24. Thunnissen, M., Boselie, P., & Fruytier, D. (2013). A review of talent management: Infancy or adolescence? *International Journal of Human Resource Management, 24*(9), 1744–1761.
25. Collings, D. G., & Scullion, H. (2012). Global staffing: A critical review. In G. K. Stahl & I. Björkman (Eds.), *Handbook of international human resource management* (pp. 141–157). London, UK: Edward Elgar.

26. Cascio, W. F., & Boudreau, J. W. (2015). The search for global competence: From international HR to talent management. *Journal of World Business, 51*(1), 103–114.
27. Lawler, E. E., III, & Boudreau, J. W. (2015). *Global trends in human resource management: A twenty-year analysis*. Stanford CA: Stanford University Press.
28. Caligiuri, P., Collings, D., & Scullion, H. (2010). Global talent management. *Journal of World Business, 43*(2), 105–108.
29. Heinecke, P. (2011). *Success factors of regional strategies for multinational corporations: Appropriate degrees of management autonomy and product adaptation*. Berlin, Germany: Springer.
30. Sparrow, P. R., & Makram, H. (2015). What is the value of talent management? Building value-driven processes within a talent management architecture. *Human Resource Management Review, 25*(3), 249–263.
31. Sparrow, P. R., Scullion, H., & Tarique, I. (2014). Multiple lenses on talent management: Definitions and contours of the field. In P. R. Sparrow, H. Scullion, & I. Tarique (Eds.), *Strategic talent management: Contemporary issues in international context* (pp. 36–70). Cambridge, UK: Cambridge University Press.
32. Garrow, V., & Hirsh, W. (2008). Talent management: Issues of focus and fit. *Public Personnel Management, 37*(4), 389–402.
33. Campbell, V., & Hirsh, W. (2013). *Talent management: A four-step approach.* Report 502, Institute for Employment Studies. London, UK: IES. Retrieved from http://www.employment-studies.co.uk/system/files/resources/files/502.pdf
34. Dries, N., Cotton, R. D., Bagdadli, S., & de Oliveira, M. Z. (2014). HR Directors' understanding of 'talent': A cross-cultural study. In A. Al Ariss (Ed.), *Global talent management: Challenges, strategies, and opportunities* (pp 15–28). New York, NY: Springer.
35. Iles, P., Chuai, X., & Preece, D. (2010). Talent management and HRM in multinational companies in Beijing: Definitions, differences and drivers. *Journal of World Business, 45*(2), 179–189.
36. Swailes, S., & Downs, Y. (2014). *Inclusive talent management: Model building and theoretical underpinning.* Working paper delivered at Human Resource Development Conference, June 4–6, 2014, Edinburgh Napier University, Scotland. Retrieved from http://www.ufhrd.co.uk/wordpress/wp-content/uploads/2014/11/Stephen-Swailes.pdf
37. Downs, Y., & Swailes, S. (2013). A capability approach to organizational talent management. *Human Resource Development International, 16*(3), 267–281.
38. Ulrich, D., & Smallwood, N. (2012). What is talent? *Leader to Leader, 63,* 55–61.
39. Collings, D. G., & Mellahi, K. (2009). Strategic talent management: A review and research agenda. *Human Resource Management Review, 19*(4), 304–313.
40. Huselid, M. A., & Becker, B. E. (2011). Bridging micro and macro domains: Workforce differentiation and strategic human resource management. *Journal of Management, 37*(2), 395–403.
41. Martin, J., & Schmidt, C. (2010). How to keep your top talent. *Harvard Business Review, 88*(5), 54–61.
42. Kantrowitz, T. M. (2014). *Global assessment trends report.* Thames Ditton, UK: CBE (Corporate Executive Board).
43. Edwards, T., Sanchez-Mangas, R., Bélanger, J., & McDonnell, A. (2015). Why are some subsidiaries of multinationals the source of novel practices while others are not? National, corporate and functional influences. *British Journal of Management, 26*(2), 146–162.
44. Cantwell, J., & Zhang, Y. (2009). The innovative multinational firm: The dispersion of creativity, and its implications for the firm and for world development. In S. Collinson & G. Morgan (Eds.), *Images of the multinational firm* (pp. 45–46). Chichester, UK: Wiley.
45. Collings, D. G. (2014). Integrating global mobility and global talent management: Exploring the challenges and strategic opportunities. *Journal of World Business, 49*(2), 253–261.
46. Minbaeva, D., & Collings, D. G. (2013). Seven myths of global talent management. *International Journal of Human Resource Management, 24*(9), 1762–1776.

47. Bidwell, M. J., & Keller, J. R. (2014). Within or without? How firms combine internal and external labor markets to fill jobs. *Academy of Management Journal, 57*(4), 1035–1055.
48. Bidwell, M. J. (2011). Paying more to get less: Specific skills, incomplete information and the effects of external hiring. *Administrative Science Quarterly, 56*(3), 369–407.
49. Dokko, G., Wilk, S. L., & Rothbard, N. P. (2009). Unpacking prior experience: How career history affects job performance. *Organizational Science, 20*(1), 51–68.
50. Cappelli, P., & Keller, J. R. (2013). Classifying work in the new economy. *Academy of Management Review, 38*(4), 575–596.
51. Cappeli, P., & Keller, J. R. (2013). A study of the extent and potential causes of alternative employment arrangement. *Industrial and Labor Relations Review, 66*(4), 874–901.
52. Osnowitz, D. (2010). *Freelancing expertise: Contract professionals in the new economy*. Ithaca, NY: ILR Press.
53. Lepak, D. P., & Snell, S. A. (1999). The human resource architecture: Toward a theory of human capital allocation and development. *Academy of Management Review, 24*(1), 31–48.
54. Becker, B. E., & Huselid, M. A. (2006). Strategic human resources management: Where do we go from here? *Journal of Management, 32*(6), 898–925.
55. Campion, M. A., Fink, A. A., Ruggeberg, B. J., Carr, L., Phillips, G. M., & Odman, R. B. (2011). Doing competencies well: Best practices in competency modeling. *Personnel Psychology, 64*(1), 225–262.
56. Guion, R. M., & Highhouse, S. (2006). *Essentials of personnel assessment and selection*. Mahwah, NJ: Erlbaum.
57. Sanchez, J., & Levine, E. (2012). The rise and fall of job analysis and the future of work analysis. *Annual Review of Psychology, 63*, 397–425.
58. Schuler, R. S. (1992). Strategic human resources management: Linking the people with the strategic needs of the business. *Organizational Dynamics, 21*(1), 18–32.
59. Meyers, M. C., & van Woerkom, M. (2014). The influence of underlying philosophies on talent management: Theory, implications for practice, and research agenda. *Journal of World Business, 49*(2), 192–203.
60. Monks, K., Kelly, G., Conway, E., Flood, P., Truss, K., & Hannon, E. (2013). Understanding how HR systems work: The role of HR philosophy and HR processes. *Human Resource Management Journal, 23*(4), 379–395.
61. Ruona, W. E. A., & Lynham, S. A. (2004). A philosophical framework for thought and practice in human resource development. *Human Resource Development International, 7*(2), 151–164.
62. Brewster, C., Gollan, P. J., & Wright, P. M. (2013). Guest editors' note: Human resource management and the line. *Human Resource Management, 52*(6), 829–838.
63. Bowen, D. E., & Ostroff, C. (2004). Understanding HRM-firm performance linkages: The role of the 'strength' of the HRM system. *Academy of Management Review, 29*(2), 203–221.
64. Ostroff, C., & Bowen, D. E. (2016). Reflections on the 2014 decade award: Is there strength in the construct of HR system strength? *Academy of Management Review, 41*(2), 196–214.
65. Baaij, M. C., Mom, T. J. M., Van den Bosch, F. A. J., & Volberda, H. W. (2015). Why do multinational corporations relocate core parts of their corporate headquarters abroad? *Long Range Planning, 48*(1), 46–58.
66. Lewin, A. Y., Massini, S., & Peeters, C. (2009). Why are companies offshoring innovation? The emerging global race for talent. *Journal of International Business Studies, 40*(6), 901–925.
67. Cerdin, J.-L., & Brewster, C. (2014). Talent management and expatriation: Bridging two streams of research and practice. *Journal of World Business, 49*(2), 245–252.
68. Ahlvik, C., & Bjorkman, I. (2015). Towards explaining subsidiary implementation, integration, and internalization of MNC headquarters HRM practices. *International Business Review, 24*(3), 497–505.
69. Ahlvik, C., Smale, A., & Sumelius, J. (2016). Aligning corporate transfer intentions and subsidiary HRM practice implementation in multinational corporations. *Journal of World Business, 51*(3), 343–355.

70. Caligiuri, P., & Bonache, J. (2016). Evolving and enduring challenges in global mobility. *Journal of World Business, 51*(1), 127–141.
71. D'Andrea, A., & Gray, B. (2013). The work and life of corporate expatriates: New patterns and regimes of mobility in the knowledge economy. *International Review of Social Research, 3*(1), 87–107.
72. Mayrhofer, W., Sparrow, P. R., & Zimmermann, A. (2008). Modern forms of international working. In M. Dickmann, C. Brewster, & P. R. Sparrow (Eds.), *International human resource management: Contemporary issues in Europe* (pp. 219–239). London, UK: Routledge.
73. Starr, T. L., & Currie, G. (2009). Out of sight but still in the picture: Short-term international assignments and the influential role of family. *International Journal of Human Resource Management, 20*(6), 1421–1438.
74. Collings, D. G., Scullion, H., & Morley, M. J. (2007). Changing patterns of global staffing in the multinational enterprise: Challenges to the conventional expatriate assignment and emerging alternatives. *Journal of World Business, 42*(2), 198–213.
75. Meyskens, M., Von Glinow, M. A., Werther, W. B., Jr., & Clarke, L. (2009). The paradox of international talent: Alternative forms of international assignments. *International Journal of Human Resource Management, 20*(6), 1439–1450.
76. PricewaterhouseCooper [PwC]. (2012). *Talent mobility—2020 and beyond*. Retrieved from http://www.pwc.com/gx/en/managing-tomorrows-people/future-of-work/pdf/pwc-talent-mobility-2020.pdf
77. Collings, D. G., McDonnell, A., Gunnigle, P., & Lavelle, J. (2010). Swimming against the tide: Outward staffing flows from multinational subsidiaries. *Human Resource Management, 49*(4), 575–598.
78. Gertsen, M. C., & Søderberg, A.-M. (2012). Inpatriation in a globalising MNC: Knowledge exchange and translation of corporate culture. *European Journal of International Management, 6*(1), 29–44.
79. Reiche, S., Kraimer, M., & Harzing, A. W. (2009). Inpatriates as agents of cross-unit knowledge flows in multinational corporations. In P. Sparrow (Ed.), *Handbook of international human resource management: Integrating people, process and context* (pp. 151–170). Chichester, UK: Wiley.
80. Froese, F. J., Kim, K., & Eng, A. (2016). Language, cultural intelligence, and inpatriate turnover intentions: Leveraging values in multinational corporations through inpatriates. *Management International Review, 56*(2), 283–301.
81. Harvey, M., Reiche, B. S., & Moeller, M. (2011). Developing effective global relationships through staffing with inpatriate managers: The role of interpersonal trust. *Journal of International Management, 17*(2), 150–161.
82. Maley, J., Moeller, M., & Harvey, M. (2015). Strategic inpatriate acculturation: A stress perspective. *International Journal of Intercultural Relations, 49*, 308–321.
83. Harvey, M., Kiessling, T., & Moeller, M. (2011). Globalization and the inward flow of immigrants: Issues associated with the inpatriation of global managers. *Human Resource Development Quarterly, 22*(2), 177–194.
84. Crowley-Henry, M., & Heaslip, G. (2014). Short-term international assignments: Military perspectives and implications for international human resource management. *European Management Journal, 32*(5), 752–760.
85. Vance, C. M., & Paik, Y. (2015). *Managing a global workforce: Challenges and opportunities in international human resource management* (2nd ed.). New York, NY: Routledge.
86. Watson-Manheim, M. B., Chudoba, K., & Crowston, K. (2002). Discontinuities and continuities: A new way to understand virtual work. *Information, Technology and People, 15*(3), 191–209.
87. Chudoba, K. M., Wynn, E., Lu, M., & Watson-Manheim, M. B. (2005). How virtual are we? Measuring virtuality and understanding its impact in a global organization. *Information Systems Journal, 15*(4), 279–306.

88. Cummings, J. N., Espinosa, J. A., & Pickering, C. K. (2009). Crossing spatial and temporal boundaries in globally distributed projects: A relational model of coordination delay. *Systems Research, 30*(3), 420–439.
89. Klitmøller, A., & Lauring, J. (2013). When global virtual teams share knowledge: Media richness, cultural difference and language commonality. *Journal of World Business, 48*(3), 398–406.
90. Shachaf, P. (2008). Cultural diversity and information and communication technology impacts on virtual teams. *Information & Management, 45*(2), 131–142.
91. Hofstede, G. (2001). *Culture's consequences: Comparing values, behaviors, institutions, and organizations across nations* (2nd ed.). London, UK: Sage.
92. Li, W. (2010). Virtual knowledge sharing in a cross cultural context. *Journal of Knowledge Management, 14*(1), 38–50.
93. Stahl, G., Maznevski, M. L., Voght, A., & Jonsen, K. (2010). Unraveling the effects of cultural diversity in teams: A meta-analysis of research in multicultural work groups. *Journal of International Business Studies, 41*(4), 690–709.
94. Gibson, C. B., Huang, L., Kirkman, B. L., & Shapiro, D. L. (2014). Where global and virtual meet: The value of examining the intersection of these elements in twenty-first century teams. *Annual Review of Organizational Psychology and Organizational Behavior, 1,* 217–244.
95. Kirkman, B. L., Shapiro, D. L., Lu, S., & McGurring, D. P. (2016). Culture and teams. *Current Opinion in Psychology, 8,* 137–142.
96. Gibson, C. B., McDaniel, D., & Szkudlarek, B. (2012). Tales from the (multicultural) field: Toward developing research conducive to proximal theory building. In A. M. Ryan, F. L. Oswald, & F. T. L. Leong (Eds.), *Conducting multinational research projects in organizational psychology: Challenges and opportunities* (pp. 9–28). Washington, DC: American Psychological Association.
97. Van den Born, F., & Peltokorpi, V. (2010). Language policies and communication in multinational companies: Alignment with strategic orientation and human resource management. *Journal of Business Communication, 47*(2), 97–118.
98. Lauring, J., & Selmer, J. (2010). Multicultural organizations: Common language and group cohesiveness. *International Journal of Cross Cultural Management, 10*(3), 267–284.
99. Lauring, J., & Selmer, J. (2011). Multicultural organizations: Common language, knowledge sharing and performance. *Personnel Review, 40*(3), 324–343.
100. Harvey, M., Novicevic, M. M., & Garrison, G. (2005). Global virtual teams: A human resource capital architecture. *International Journal of Human Resource Management, 16*(9), 1583–1599.
101. Zey, M. G. (2012). Virtual teams: The problems and possibilities of computer and cyber-based global work groups in modern organizations. *Journal of International Management Studies, 12*(3), 1–23.
102. Caligiuri, P., Lazarova, M., & Tarique, I. (2005). Training, learning, and development in multinational organizations. In H. Scullion & M. Linehan (Eds.), *International human resources management: A critical text* (pp. 71–90). New York, NY: Palgrave Macmillan.
103. Ko, H.-C., & Yang, M.-L. (2011). The effects of cross-cultural training on expatriate assignments. *Intercultural Communication Studies, 20*(1), 158–174.
104. Littrell, L., & Salas, E. (2005). A review of cross-cultural training: Best practices, guidelines, and research needs. *Human Resources Development Review, 4*(3), 305–334.
105. Schneider, B., Ehrhart, M. G., & Macey, W. H. (2013). Climate and culture. *Annual Review of Psychology, 64,* 361–388.
106. Osland, J. S., & Bird, A. (2000). Beyond sophisticated stereotyping: Cultural sensemaking in context. *Academy of Management Executive, 14*(1), 65–77.
107. Gould, S. J., & Grein, A. F. (2008). Think globally, act globally: A culture-centric comment on Leung, Bhagat, Buchan, Erez, and Gibson (2005). *Journal of International Business Studies, 40*(2), 237–254.

108. Gibson, C. B., & McDaniel, D. (2010). Moving beyond conventional wisdom: Advancements in cross-cultural organizational behavior theory. *Perspectives on Psychological Science, 5*(4), 450–462.
109. Tatoglu, E., Glaister, A. J., & Demirbag, M. (2016). Talent management motives and practices in an emerging market: A comparison between MNEs and local firms. *Journal of World Business, 51*(2), 278–293.
110. Hartmann, E., Feisel, E., & Schober, H. (2010). Talent management of western MNCs in China: Balancing global integration and local responsiveness. *Journal of World Business, 45* (2), 169–178.

What Is Talent Management? The Perception from International Human Resources Management Students

João Leite Ribeiro and Delfina Gomes

Abstract Organisations are facing important challenges in what concerns human capital, and one of these challenges is talent management. Different reasons have contributed to highlight the importance of talent management, mainly due to the competition between employers at the local level but even more at the global level. The discourse within Human Resources Management (HRM) highlights the maximization of the talent of individual employees as a unique source of competitive advantage for organisations. The main objective of this chapter is to contribute to a better understanding of the meaning attributed to the concepts "talent" and "talent management", from the point of view of students of the curricular unit International Human Resources Management. More specifically, the study conducted aims to understand how HRM may play a relevant role in attracting and retaining talents.

1 Introduction

Twenty-first century organisations are facing important challenges in what concerns human capital. One of these challenges is talent management [1; see also 2]. In fact, different reasons have contributed to highlight the importance of talent management, mainly due to the competition between employers at the local level but even more at the global level [3; see also 2, 4]. According to Farndale et al. [3, p. 161; see also 5], "[t]his globalization of talent management brings with it a requirement to create new HRM tools, methods and processes to provide the necessary co-ordination systems to support global integration".

J.L. Ribeiro (✉) · D. Gomes
School of Economics and Management, University of Minho, Braga, Portugal
e-mail: joser@eeg.uminho.pt

C. Machado (ed.), *Competencies and (Global) Talent Management*,
Management and Industrial Engineering, DOI 10.1007/978-3-319-53400-8_4

More recently, the discourse is the one where the central element of strategic human resources policy is the maximization of the talent of individual employees as a unique source of competitive advantage for organisations [3, 6, 7]. However, the problem that emerges is the extent to which organizations effectively manage their talent in this respect [3, 8].

Nonetheless, this is still an under-researched area, characterized by little theory development and reduced empirical evidence [1, 9]. In a context of increasingly global competition where it is recognized the importance of talent management for success, according to Farndale et al. [3, p. 161] "the concept is still lacking in definition and theoretical development".

The main objective of this chapter is to contribute to a better understanding of the meaning attributed to the concepts "talent" and "talent management", from the point of view of students of the curricular unit International Human Resources Management. More specifically, the study conducted aims to understand how Human Resources Management (HRM) may play a relevant role in attracting and retaining talents. Through the use of specific organisations, seen as exemplar in the management of talents, the participants reveal their perceptions about these concepts and the development, waste or loss of talents, as well as the role that HRM has in this subject.

This chapter is structured in 5 sections. The second section provides a literature review on the concept of talent management from different perspectives and from the perspective of HRM. The third section describes the research method adopted in this study and is followed by the analysis and discussion of the results in section four. The chapter ends with the conclusion.

2 Talent Management: An Overview of the Literature

2.1 Talent Management: Different Perspectives

According to Tarique and Schuler [10, p. 124], although there seems to be a growing consensus regarding the meaning of "talent management", when considering the global environment the exact meaning of global talent management is not so precise. The global talent management will be influenced by the context it appears in and sometimes is used interchangeably with International Human Resources Management [10; see also 11–13].

The 1990s were characterized by a high demand for talented employees, what confronted the organizations with a global shortage of talent, which surpassed the supply [14; see also 15, 16]. This shortage of talent put the focus on how to acquire, retain and manage talent in global business [14; see also 17]. This challenge came to be labelled as "global talent management" [18].

The literature has developed a wide range of perspectives to conceptualize and contextualize talent management [19, 20]. Four main perspectives co-exist [19; see 7, 9].

According to Cooke et al. [19], the first perspective is a universalist and inclusive approach to talent management, which argues that all employees have talent and that talent should be harnessed for the organizational good through a range of HRM practices. Seen as a new fashion of HRM, this perspective was criticized for being undifferentiated [19; see also 7, 21, 22].

The second perspective takes a narrow view and defines talent management as succession planning. In this approach, according to Cooke et al. [19, p. 226]:

> … a key task is to develop 'talent pipelines' to ensure the current and future supply of employee competence, as well as an organization-wide, holistic talent mindset [7]. Underlining this perspective is a long-term and static view that assumes that what is required in the future (i.e. roles and persons for the roles) is known to the organization, and that what the organization needs to do is to plan for it.

The main critic to this perspective is that it failed to take into account business and labour market uncertainties [8, 19, 23].

The third perspective defines talent management as the management of only relatively small group of employees who demonstrate considerable potential, that is the management of talented employees [19; see also 16, 21, 22]. Managing only talented employees implies to identify who those persons are through pre-defined criteria and "then manage them effectively through a set of tightly coupled HRM tools, activities, and processes" [19, p. 226]. However, the focus only on individuals identified as talented may be negative to the organizational culture, discouraging teamwork and collaborative spirit [19].

The fourth approach moves from individuals to strategic positions, where talent management is the "strategic management of 'pivotal positions' rather than 'pivotal people'" [19, p. 226]. According to Cooke et al. [19, p. 226]:

> Building on the works of Huselid et al. [24] and Boudreau and Ramstad (2005), which argued for an increased focus on key positions instead of talented individuals, this perspective of TM [Talent Management] focuses on organizational processes and systems for identifying key positions that are strategically important to the organization and filling them with the right personnel through well-developed HR systems and processes [9]. These key positions are not confined to managerial roles, and may include functional and technical positions, which may have a significant impact on organizational performance [9].

2.2 *Talent Management: HRM Definitions*

The four main perspectives used to conceptualize and contextualize talent management are the result of different definitions of talent management that can be found in HRM literature, as presented by Dries [1, p. 274], as follows in Table 1.

However, according to Dries [1], most of the authors fail to provide a definition of what is "talent" or what type of practices are under the talent management label.

Table 1 Definitions of talent management in the HRM literature

Reference	Definition of talent management
Sloan et al. [33, p. 236]	"Managing leadership talent strategically, to put the right person in the right place at the right time"
Pascal [43, p. 9]	"Talent management encompasses managing the supply, demand, and flow of talent through the human capital engine"
Ashton and Morton [2, p. 30]	"TM is a strategic and holistic approach to both HR and business planning or a new route to organizational effectiveness. This improves the performance and the potential of people—the talent—who can make a measurable difference to the organization now and in future. And it aspires to yield enhanced performance among all levels in the workforce, thus allowing everyone to reach his/her potential, no matter what that might be"
Duttagupta [44, p. 2]	"In the broadest possible terms, TM is the strategic management of the flow of talent through an organization. Its purpose is to assure that a supply of talent is available to align the right people with the right jobs at the right time based on strategic business objectives"
Warren [45, p. 26]	"In its broadest sense, the term can be seen as the identification, development, engagement, retention and deployment of talent, although it is often used more narrowly to describe the short- and longer-term resourcing of senior executives and high performers"
Jerusalim and Hausdorf [46, p. 934]	"High potential identification and development (also known as talent management) refers to the process by which an organization identifies and develops employees who are potentially able to move into leadership roles sometime in the future"
Capelli [8, p. 1]	"At its heart, talent management is simply a matter of anticipating the need for human capital and setting out a plan to meet it"
Collings and Mellahi [9, p. 2]	"We define strategic talent management as activities and processes that involve the systematic identification of key positions which differentially contribute to the organization's sustainable competitive advantage, the development of a talent pool of high potentials and high-performing incumbents to fill these roles, and the development of a differentiated human resource architecture to facilitate filling these positions with competent incumbents and to ensure their continued commitment to the organization"
Silzer and Dowell [36, p. 18]	"Talent management is an integrated set of processes, programs, and cultural norms in an organization designed and implemented to attract, develop, deploy, and retain talent to achieve strategic objectives and meet future business needs" (p. 18)

Source Dries [1, p. 274]

2.3 *Talent Management: Human Resources Management and Relevant Theoretical Perspectives*

There are different relevant theoretical perspectives, according to Dries [1], to the study of talent management. The perspective of human capital assumes a resource-based view of the company, where the main criterion is the contribution of employees to the organization. The focus is devoted to the human capital that is both valuable and unique in an organization. There is also the perspective of international (cross-cultural) HRM, which has emerged in recent years, focused on the global talent management. However, there is still a lack of research focusing on the experiences and perceptions of talented (and less talented) employees [see, as exceptions, 16, 25], as well as research focusing on advantaged and disadvantaged employee groups from the viewpoint of workforce differentiation [see 26].

According to Schuler, Jackson and Tarique [14, p. 506], there are a considerable number of practices that companies can use in their initiatives concerning global talent management, which include the following: location planning and relocation management; HR planning and forecasting; staffing, meaning attraction and selection, retention, reduction and removal; training and development; performance assessment; and compensation. However, "[m]atching an accurate diagnosis of a firm's strategy and talent management situation with possible HR policies and practices is a first step in gaining and sustaining a global competitive advantage that may result from the successful implementation of the appropriate HR policies and practices" [14, p. 506]. The contribution that HR professionals can make to the global talent management initiatives of a company is by measuring the impact of HR policies and practices through the use of metrics aligned with the business strategies [14]. These "[p]erformance metrics that reflect desired strategic business outcomes may include revenue, profit targets or retention of direct reports" [14, p. 506].

As expected, talent management, as HRM practices in general, is strongly influenced by the institutional and cultural context of each society and the way standard practices, such as legislation and institutional norms, are implemented, as well as the use of additional distinctive practices are the paths to firms raise competitive advantages [19].

Although the question has been raised that talent management is just a management fashion, several authors have "concluded that talent management does in fact add value to other strategic human resource management (SHRM) practices" [1, p. 274; see 22, 27, 28]. Talent management is seen by some authors as part of an organizations' HRM strategy, while others argue that "talent management is a 'mindset' and thus, an all-encompassing characteristic of an organization much like organizational culture" [1, p. 275].

In fact, it is in the context of IHRM that global talent management and its multiple potential challenges, according to Tarique and Schuler [10 p. 123], can be examined. IHRM has been challenged, during the last two decades, by several changes as follows: the introduction of increased worldwide economic development; extensive global communication; rapid transfer of new technology; growing trade; and emigration of large numbers of people [10, p. 123].

Finally, it is important to highlight, as argued by Dries [1, p. 283], that no perspective on talent management is better than the other. Regardless of best practices, in what concerns talent management the most important is organizational fit: "fit with strategic objectives, fit with organizational culture, fit with other

HR practices and policies, and fit with organizational capacity" [1, p. 283]. Additionally, there are also the implications for individual employees in what concerns talent management:

> As for individual employees, they are often oddly unaware of the talent management dynamics operating within their employing organizations—even though these are likely to have crucial implications for the further course of their career (Larsen et al. 1998). Part of the explanation is that talent management procedures are often quite intransparent, with crucial information being withheld from employees (e.g., not being identified as talented) (Bournois and Rousillon 1992). In addition, employees (even high potential ones) are often naïve, and somewhat reactive, when it comes to managing their own careers (e.g. Dries and Pepermans 2008). Advances in the academic literature may help both organizations and individual employees make more sense of how strategic talent management decisions may or may not affect them [1, p. 283].

3 Research Method

Positioned within the interpretative paradigm, this study assumes that reality is a social construction that cannot be understood independently from the actors that create that same reality [29, p. 59]. Within this paradigm, a qualitative approach is adopted in order to understand the phenomena through the meanings that individuals attribute to them [30]. Therefore, the main objective of this chapter is to contribute to a better understanding of the meaning attributed to the concepts *talent* and *talent management* and of how HRM may play a relevant role in attracting and retaining talents, from the point of view of students of the curricular unit International Human Resources Management.

Research on talent management has been mainly focused on assessing if firms adopt strategic talent management policies and practices, with researchers defining at the start what talent is; or researchers exploring how companies define talent and how they manage talented employees [19]. Given the limited studies in Portugal about HRM and talent management and the fact that this phenomenon is developed in the practice of HRM or/and by HR consulting companies, this study provides a different perspective of this topic by analyzing the perceptions of third year students that are finishing their graduation and will be facing the challenges of finding a job. Portugal has been since 2008 facing a difficult financial situation with repercussions in all domains of society. One of these was an exodus of qualified young people. Therefore, in the context of the curricular unit of International Human Resources Management, the students were invited to define what talent is and what talent management means (below the list of questions asked will be presented). In doing so, it was possible to explore how these students perceive the importance of knowing what a talent is and what perceptions they have about which companies are exemplar in the management of talents and about the practices adopted regarding the development, waste or loss of talents.

Each of the respondents are enrolled in the curricular unit where the topic of international talent management is debated and group work on the topic is developed by the students. The first author is the teacher of this curricular unit and permission was asked to the students to pass the paper version of the qualitative survey so they could fill out with their writing and without any interference. This process was developed at the beginning of the school year (September, 2016), with just a contextualization of the research. Confidentiality was guaranteed to respondents. In surveying the respondents, they were asked to address the following questions, originally written in Portuguese and analyzed in Portuguese and only after the results were translated into English:

1. What talent management means to you?
2. For you what is a talent?
3. How can HRM help a company to attract talent?
4. How can HRM help a company to retain talent?
5. Can you identify three talents (individual and/or collective)?
6. Can you identify a business company that, in your opinion, adequately manage its talents?
7. Can you identify any team/person that has done or does good talent management?
8. In your opinion, how can a company waste/lose its talents?
9. Thinking about yourself, what talents do you have?

The curricular unit is optional and students from a diversified number of courses are enrolled, in a total of 50 Portuguese students, from Management (19), Economics (3), International Relations (26), and International Business (2). Most of the students (43) age ranges between 20 and 22 years old, with 22 men and 28 women. The average time of respondents was 40 min and all questions were answered. From the 50 students, only the ones from management had previously attended a curricular unit of HRM. Content analysis was adopted following a thematic line guided by the research questions that were reflected in the survey questions, according to the list provided above.

4 Findings and Discussion

In this section, the findings and their discussion are presented following the thematic line guided by the research questions, as mentioned above. The quotations from the respondents are classified as a number from 1 to 50, followed by M or F, which means masculine or feminine. To simplify at the end of each quotation appears subject, the number and gender, for example Subj. 2F.

4.1 What Talent Management Means to You?

The results show that the respondents adopted different dimensions to define talent management, as can be identified in the literature [1, 9, 16]. From the point of view of responsibility, the respondents have highlighted corporate responsibility and individual responsibility. In what concerns corporate responsibility, the company is considered responsible for identifying the talents that it has inside, in its workforce. It is considered to be fundamental that organizational structures understand that persons are not only good in some characteristics, "… but with persistency and good management capacity and leadership other personal skills can be improved and even new talents may be discovered within the workforce already in the company" (Subject 4F; with more 29 similar answers).

When considering individual responsibility, the respondents argue that the individual has to be proactive and create the conditions for his own development and progress, "not waiting for others to take care of my talent" (Subj. 1M), with more 32 similar answers). Developing a strategy of self-knowledge is fundamental. As described by Subj. 2F: "It is important to know your talents, to develop them and to empower them in contexts where they are most profitable to the individual and also where they allow to bring more benefits to companies where the individual can develop his activity and be recognized by it" (with more 22 similar answers).

This responsibility for identifying and developing talent, both as personal, individual or group characteristics, is described in the literature in the double perspective—corporative and individual, with studies that highlight this shared responsibility [1].

Another perspective adopted to explain what talent management is the aim: corporate aim and individual aim. From the perspective of corporate aim, talent management is intended to obtaining added value, greater profitability and a utilitarian orientation of the people, in line with Collings and Mellahi [9]. In the words of one of the respondents, talent management is: "When you recognize a talent and a strategy and dynamic is developed so that this talent reaches its maximum productivity, that is, to explore that talent and know how to manage it that it can then be applied to different areas and situations" (Subj. 25F; with more 26 similar answers). From the individual aim, talent management is related to obtaining success, personal visibility, higher and better status.

The globalization of the economy and the greater competitiveness between the organizations lead to this type of purposes that imply new strategies and new instruments of talent management [5, 9, 16, 20, 31].

Another perspective, expressed by 39 of the respondents, used to define talent management is the perspective of commitment, in which talent management is the management of people with the capacity to give to the organization the skills which bring usefulness and added value to the organization, but simultaneously, the organization must know their workers very well and apply strategies that contribute to their continuous development. In the words of the respondents:

> … discover what makes people feel fulfilled on a personal level and can positively impact the organization. It is not a static process because it requires education and the constant challenge of people and companies (Subj. 10M);

> Manage the better and more interesting a business has. Managing a talent means taking advantage of the skills that a person has and taking advantage of it in the development of the company (Subj. 35F).

> Recognising the characteristics that a person has and that makes that person different from the others in a professional context, and place that person in an area of the company where she can use her talent only to bring benefits to the company and worker himself. (Subj. 49M, but also mentioned by more 34 respondents).

To define talent management from the perspective of commitment can range from more informal aspects to issues of a formal nature at the level of strategic human resources management, as demonstrated by previous studies [3, 6, 7, 18, 32, 33].

The perspective of values is another approach to define talent management. Within this perspective, the company has "to assume itself as a serious and honest entity in the relation that establishes with its workers knowing to take advantage of the qualities of a person" (Subj. 4F; with more 28 similar answers), and "encouraging them to adopt a global and integral development as a person and as a professional" (29 respondents express this idea), in line with the literature [2, 6–8]. As argued by one of the respondents, "talent management implies an investment by the organization. It is about providing someone, who has a differentiating and advantageous characteristic for the company, with opportunities to apply his talent and provide training so that it develops even more" (Subj. 19F; with more 34 similar answers).

The analysis of the data reveals a strategy of reciprocal instrumentality [7, 17, 18]. On the one hand, "Organizations recognize, attract and manage the talents by placing them in positions where they are needed and provide added value to the organization" (Subj. 3F; with more 32 similar answers). Taking advantage of characteristics, skills and abilities, thus benefiting from their talents. On the other hand, people "should know how to use characteristics that set them apart from others, at the right time and at the right place" (Subj. 5F) and "it is not worth wasting talent if what they ask of us has nothing to do with what we stand out most" (Subj. 6F).

According to the data, the instrumental perspective of the use of talent by workers should privilege aspects of personal distinctiveness, interpersonal comparability and take on a sense of opportunity and personal utility, which is aligned with the literature [16, 19, 21, 22].

The respondents also contextualize talent management taking in consideration both workers and organizations in varied scenarios: "Not all HRs in the professional contexts in which they are inserted are talented. Talent management is to know how to evaluate and to gather in itself all the necessary capacities for a certain plan of action" (Subj. 49M); and "… coordinate, direct, choose the talents appropriate to the needs of the company, business or activity…" (Subj. 41F). This need of contextualization of what is and can mean a talent and lead to its development or disappearance is aligned with previous studies [for example, 7, 9–13, 20].

4.2 For You What Is a Talent?

The analysis of the data of the respondents allows to conclude for the absence of a specific definition of talent, appearing this concept associated not only to personal characteristics, but also to individuals, groups and even concrete organizations, reflecting the fail also identified in the literature to provide a definition of what is *talent* [9–13].

From the perspective of individual characteristics, the concept is presented as "… something intrinsic of the person and that it manages to perform almost naturally and even when it demands work it seems that it does it without any effort" (Subj. 3F). Several respondents (39 answers) consider that talent is a capacity, competence or skill that stands out most in a person and that will be used by the company.

Talent is also understood within this perspective as "… an unusual, extraordinary capacity that a person possesses and that stands out from all others in a given area" (Subj. 14M). This differentiating and unique capacity still appears "as an innate aptitude" (Subj. 31F) or "acquired by an individual" (Subj. 38M), that is "a natural or worked capacity over time, but that without a doubt will make a person stand out from others" (Subj. 39M), in line with the literature [16, 21, 22, 32].

Talent as a polysemous and polymorphic concept is equally perceived as something that can designate an individual, group or organization, as stands out from the respondents answers:

> A person or group of people who have a characteristic and/or ability that makes them stand out positively in the performance of a task from other people (Subj. 40M);

> A group of people who are the best at what they do, who are hard to compete with; Who give their maximum in what they are really good at and strive to be better and better (Subj. 20F);

> There are companies that are themselves a talent, since they function in such a way that the whole is different from the sum of the parts. That is, all the individual talents that exist are as if powered by something that can be the company's culture, values, goals or even leadership and with it the company is something even better and distinct from most other companies (Subj. 15M).

A talent can be "someone who stands out for his ability to adapt and to maintain high levels of performance even in professional areas he does not like…" (Subj. 25F), which clearly emphasizes adaptability and resilience as key to defining and building a talent [24, 34]. Finally, the respondents highlight the fact that whether at the individual, group or organizational level there is a need for dynamism, for development and continuous improvement (38 subjects provide similar answers).

4.3 How Can HRM Help a Company to Attract Talent?

At the level of how HRM can contribute to the attraction of talent, and also to retain as per the next question, there is a tendency for responses not to consider this department or direction apart from the other areas of management, but in interconnection with them, characterized by a transversality of themes, actions and diffusion of policies, and practices in the company as a whole [7, 10].

As to the factors or strategies that favour talent attraction by HRM, the respondents mention: the credibility of the company in the market (26 answers in 50), standing out at this level the "perception of honesty and seriousness of the company" (Subj. 1M), and "the level of demand associated with a reputed company, responsible at the technical, social and environmental level" (Subj. 3F).

Another category of responses highlights *recognition*, which is related not only to remuneration, incentives and benefits policies (39 answers in 50), but also to "practices that keep motivated, happy, and satisfied talents with the work they do" (Subj. 3F). As highlights by one of the respondents: "When we work in a company that shows interest and concern for its employees, it is much easier to captivate and keep their workers fulfilled" (Subj. 4F).

As regards good working conditions, these are translated by respondents not only in terms of physical conditions, but also in what they call "good working environment for relations between people in different sectors" (Subj. 2F), and the potential for growth and development through psychological and mental conditions to perform work, in this case to express talent. For example: "Have a team work together, show the importance that this talent represents. Give freedom to this talent to suggest new procedures or necessary changes. Give the necessary support so that his function is fulfilled successfully" (Subj. 5F).

More specifically and explicitly referring to the performance of HRM within the scope of talent attraction, the following aspects stand out from the data, as Table 2.

4.4 How Can HRM Help a Company to Retain Talent?

Regarding the retention of talent, as also present in the study by Ashton and Morton [2], one of the ideas mentioned by 39 respondents can be translated into:

> When we work in a company that shows interest and concern for its employees, it is much easier attract and to keep employees motivated. Although this aspect is the responsibility of the entire organizational structure, it seems to me that it should be led by the HRM area (Subj. 4F).

Respondents also refer to the need to encourage, recognize and effectively value the worker considered a talent, so that the worker feels useful and a person properly integrated and "…not a rare bird but a person" (Subj. 20F). According to other respondents:

Table 2 Performance of HRM within the scope of talent attraction

HRM action	Respondents
The ability of the HRM body to generate and promote feelings of belonging	32 respondents
The ability to "diagnose the potential of people, prepare and instruct them, creating the conditions for the development of their potential and their own career" (Subj. 6F)	35 respondents
"Capacity and ability to develop policies that recognize merit, create career development opportunities within the company and foster mutual trust" (Subj. 13M)	30 respondents
Provide and promote the development of contexts that allow good use and improvement of individual and group talents	24 respondents
"To promote policies and practices that contribute to the development not only of the professional but of the person" (Subj. 10M)	39 respondents
Have an attractive communication ability, showing that they have a good working environment and that people in HRM are responsible and interested in employees	29 respondents

> Talent lacks a permanent game to play and it is up to HRM to monitor and stimulate areas of the company through direct leadership to provide challenging play. There must not be any discrimination of any kind, particularly because it is a talent, but good integration policies and practices in which the person feels that he is contributing to the company and is encouraged and recognized (Subj. 21F).

> Develop your potential without leading to exhaustion and providing personal development and satisfaction at the same time (Subj. 32F).

> It is essential to make the "Talent" understand that in addition to working for the organization he is also working for himself, as if he himself was a brand and over time some advantages accompany the evolution of talent (Subj. 39M).

According to respondents, policies and practices of career management are considered fundamental to retain the talents (33 respondents). Additionally, respondents highlight that HRM has a particular responsibility of HRM, in articulation and sustained by the top managers of the organization, in the following policies: continuous training of all workers and good working conditions (36 Subj.); fair and attractive practices of remuneration, incentives and benefits (39 Subj.); and stimulating work tasks (30 references), as identified in the literature [9, 35]. According to the respondents:

> To be in constant evolution so that workers feel that they always have something new to learn and do not feel unmotivated (Subj. 36M).

> Improving working conditions and providing important positions and status appropriate to the contributions coming from these talents (Subj. 38M).

Another aspect emphasized by the respondents has to do with the prospect that HRM should be a "trustworthy and coherent area between what it claims to be and do and what it actually does" within management (42 Subj.), what is in line with the literature on the distances between the narratives resulting from the rhetoric and the daily reality that many people face [1, 6–8, 36].

4.5 Can You Identify Three Talents (Individual and/or Collective)?

When asked to identify talents, respondents diverged in the characteristics and nature of what is a talent. The answers can be classified in three categories: personality, knowledge and group, as presented in Table 3, with the number of references (refs.) made by the respondents.

A different categorization also emerged when talent was projected as an individual or an organization. Table 4 presents the top 5 talents identified by the respondents for each category.

These categories contain a set of characteristics, competencies and capacities of diverse nature that define a talent and that allow to perceive it as an important resource for the organization. A resource with unique value, which allows in theoretical terms an approach to resource-based theory [37]. Based on this theory, talent can be perceived as a resource that is an asset, which will add value to the organization and mark it in a distinctive way [16, 25, 37].

The respondents also adopted a perspective of talent as a natural gift, with the most important 10 being: communication (27 refs.); drawing (15 refs.); painting (13 refs.); dance (12 refs.); resolving conflicts (10 refs.); writing (7 refs.); musical skill (16 refs.); sports fitness (14 refs.); leadership (16 refs.); persistence and overcoming (22 refs.).

Table 3 Categories and talents

Categories	Talents
Personality	Persistence (9 refs.) Resilience (8 refs.) Responsibility (8 refs.) Creativity (7 refs.) Critical spirit (2 refs.) Humility (1 refs.) Persuasion (1 refs.)
Knowledge	Know how to communicate (12 refs.) Leadership ability (7 refs.) Global management capacity (1 refs.) Organization (1 refs.) Reasoning ability (1 refs.) Conflict management (1 refs.)
Group	Teamwork (7 refs.) Proactivity (4 refs.) Work capacity (2 refs.)

Table 4 Individuals and organizations as talents

Categories	Talents
Individuals	Cristiano Ronaldo (42 refs.) Pope Francis (31 refs.) Freddy Mercury (24 refs.) Steve Jobs (29 refs.) Picasso (28 refs.) Clint Eastwood (22 refs.) John Lennon (19 refs.) Barack Obama (17 refs.) Meryl Streep (13 refs.) Leonard da Vinci (8 refs.)
Organizations	Portuguese Football Team (34 refs.) Google Inc. (26 refs.) Apple Inc. (24 refs.) Leicester City Football Club (18 refs.) Continental AG (Portuguese subsidiary company) (16 refs.) AIESEC (international students association) (15 refs.) BBC Station (14 refs.) International Amnisty (12 refs.) United Nations (11 refs.) Unicef (8 refs.)

Table 5 Business companies that adequately manage their talents

Portuguese companies	Non-Portuguese companies
SONAE Farfetch Primavera Business Software Solutions Logoplaste Jerónimo Martins Continental AG (Portuguese subsidiary)	IKEA International Group (Sweden) Apple Inc. (EUA) Microsfft Corporation (EUA) Walt Disney World Resort (EUA) Google Inc. (EUA) Lidl & Cia (Germany) Facebook Inc. (EUA) FNAC (France) BBC (United Kingdom) COCA-COLA (USA) Inditex (Spain)

4.6 Can You Identify a Business Company that, in Your Opinion, Adequately Manage Its Talents?

The fact that the respondents are Portuguese nationals will have determined the reference to Portuguese business realities. The twenty organizations most referenced as having HRM practices best suited for talent management, are as presented in Table 5.

These organizations are considered by the respondents as having appropriate leaderships that promote a culture of development of their people [27, 38], and an HRM area with distinctive policies and practices in the human resource management, recruitment and selection, training and development, performance evaluation and compensation systems, in line with the literature [14, 18, 33].

Participants also refer to the importance of coherence of HRM policies and practices among themselves and with company culture and values, an assertion that may correspond to the vertical and horizontal fit already discussed in the literature [17, 39, 40].

4.7 In Your Opinion, How Can a Company Waste/Lose Its Talents?

The respondents, when answering to this question, also contextualize *talent* in three perspectives: as concrete people; as cognitive, practical or behavioural characteristics (that is, technical, practical or behavioural competences); as a natural gift or result of a learning process, in both cases always lacking in development.

Respondents consider that the major cause of wasteful/lost talent has to do with poor organization management and poor human resource management. This poor management of human resources is a consequence of a devaluation of people considered talents or of the skills that characterize them as talents (39 refs.). But also it is a result of the ignorance of the people who compose the company, which leads to a waste of the talent and the value of the people and consequently to the demotivation and exit of the company (32 refs.), as identified in the literature [14, 18, 33, 40). As one of the respondents says: "Failure to value or help your employees develop and feel satisfied, associated with poor management, can be a great way to lose talent and people" (Subj. 1M).

It is also mentioned that "the ignorance of people and their characteristics by companies and even by direct leaders" (Subj. 4F; with more 36 refs.), along with the lack of effective and affective recognition leads to perceptions of injustice, feelings of devaluation and, in some cases, to the loss of positive self-esteem, may result in the loss of key people or skills in many companies (42 refs.). Other aspects that lead to wasting/losing talent in respondents' perspectives are as follows: placement in areas that do not allow the application of talent or that have little to do with the potential use and development of talent (28 refs.) or areas with which people do not identify (37 refs.); policies of discrimination and "organizational cynicism" that destroys the will to progress (Subj. 23M, plus 24 refs.). One of the respondents says.

> A company can lose or lose its talents if it does not know them in the first place, and often even the direct bosses do not know their own people well. Secondly, people are placed in functions, areas or assume responsibilities with which they do not identify themselves, do not take advantage of their capacities and often are not well integrated and the most normal is to become discouraged… (Subj. 25F).

Additional reasons are presented by the respondents: Bad evaluation of workers (21 refs.) "often turning people into machines and treating them as machines" (Subj. 29F); Failure to comply with the conditions agreed upon during the selection process or after joining the company, leading to a breach of confidence (14 refs.), the non-material and psychological recognition (17 refs.). In this regard, it is further noted that "if there is no compatibility between talent and the company, if the talents are not satisfied and the company does not seek to meet the needs and expectations through coherent and fair HRM practices it is very likely that people become unmotivated and end up leaving…" (Subj. 50F, plus 19 refs.) [see 9, 17, 39, 40]. These causes are according to the participants responsible for waste/loss of talent. However, a significant number of respondents share the argument that the waste of talent can be very damaging to the company in terms of reputation, reputation, results and sustainability in the medium and long term (44 refs.) [3, 9, 14, 24].

4.8 *Can You Identify Any Team/Person that Has Done or Does Good Talent Management?*

In order to explore other entities, and not limit the analysis of talent management to organizations or corporate businesses, the respondents were asked to identify teams/persons recognized by good practices of talent management. The results are synthesized in Table 6. The three categories along which the answers can be structured are: sport teams, organizations and individual entities.

Within the category of sport teams, the main aspects highlighted were the work carried out in the field of training and development of talents both individually, where the team has players that are clearly identified as playing the central role, but

Table 6 Teams/persons with good talent management

Category	Teams/persons
Sport Teams	Sport Lisboa e Benfica (Portugal) Sporting Club de Portugal (Portugal) Portugal National Football Team (Portugal) AFC Ajax (The Netherlands) Futbol Club Barcelona (Spain) Real Madrid Club de Fútbol (Spain) Barcelona Basketball Team (Spain) Leicester City Football Club (United Kingdom)
Organizations	NASA (EUA) UNESCO (France) AIESEC (Netherlands) Amnesty International (United Kingdom) UNICEF (New York City) United Nations (New York City)
Individual entities	Companies CEO's Presidents of sport clubs Presidents of International Organizations Leonard Cohen Green Day (USA)

also the talent of the group and not of individuals. This situation happened not only with the Portuguese Football Team that won the European Championship in 2016, but also with Leicester City that won the English Premier League (year 2015–2016) without any "football star", contrary to most expectations, especially of bettors. In what concerns organizations, the main aspect is related with how they attract talent, form and commit to their retention.

The identifications of individual entities considered to have the capacity to develop sports and business teams focused on the coaches of the mentioned football teams and the CEOs of the companies mentioned in the previous question, as well as presidents of the organizations also previously referred. Of note is the reference to Green Day and the singer Leonard Cohen and their musicians, as they are considered as examples in managing their own talents, as well as the talent of others with whom they work, and being good examples in the mastery of the arts, the art of managing talents.

It is also worth mentioning that of the 50 participants, 27 also referred to their families as successful team cases, assigning an important role to parents, and particularly grandparents, "in the integrated and sustained consolidation and development of the family that is a reality and a group of high complexity" (Subj. 2F). It should be recalled that these results cannot be separated from the fact that the respondents are Portuguese, young students completing their undergraduate course.

4.9 Thinking About Yourself, What Talents Do You Have?

The respondents were asked to say what their main talents were. The tendency of the answers obtained can be classified in a perspective of greater operationalization at the level of knowledge, namely cognitive, practical, behavioural and evolutional, as per Table 7.

At this level, it is not easy to establish boundaries and categories differentiated by the four types of knowledge. There is a transversality between them and a diversity of meanings that each one of these talents can assume. It is also worth noting that some of the talents are oriented more to processes, others to achieve results and others to aspects of interpersonal relationship.

Globally, the talents that have most predominated were the ability and willingness to learn, leadership, ability to work as a team, creativity and persistence, tolerance and willingness to help others. To these are also added the so-called natural gifts: playing soccer, being a good student, singing, dancing, creative writing, writing ability and oratory speech and sense of humour.

Finally, only two participants reported not being able to identify their talents and they even consider that what others designate and assume as talents are not recognized by them as sufficiently distinctive to be understood as a talent.

Table 7 Main talents of the respondents

Perspective	Talents
Cognitive	Language proficiency Communication skills Conflict resolution skills Planning and organization skills Strategic vision Creativity Deduction ability
Practical	Ability to manage in risk situations Work capacity Quick decision-making Problem-solving ability Pragmatism Multitask Discernment in the approach and analysis of situations
Behavioural	Patience Persistence Tolerance Open mind Sympathy Understanding and assertiveness Leadership Willingness to help and work as a team Ease of communication Dynamism
Evolutional	Learning ability Desire to learn Ability to learn fast Develop critical sense Strive to improve Perseverance Commitment Capacity for innovation Resilience

5 Conclusion

This chapter aimed at contributing to a better understanding of the concepts of "talent" and "talent management", from a different perspective of this topic by analyzing the perceptions of third year students that are finishing their graduation and will be facing the challenges of finding a job. Therefore, in the context of the curricular unit of International Human Resources Management, the students were invited to answer a set of questions to explore how they define talent and talent management. Based on the obtained answers, it was possible to explore how these students perceive the importance of knowing what a talent is and what perceptions they have about which companies are exemplar in the management of talents, as well as the role that HRM can play in attracting, retaining or wasting/losing talents.

Aligned with the literature the results show that it was not easy for the students to define what talent management is. Different dimension were adopted by the respondents, with a dichotomy on corporate responsibility and individual responsibility for identifying and developing talent. This dichotomy was complemented not only with another corporate versus individual aim, but also with the concepts/strategies of commitment, values and reciprocal instrumentality. Interesting is that talent management, from the different dimensions through which it was defined, is considered to be a responsibility of the organization as well as of the individual.

When trying to define what talent is, the respondents fail in their attempt, as in previous studies 8 [1]. In their attempt, students have related the concept with different characteristics at the personal, individual, group and organizational levels. But most importantly, respondents highlight that, regardless of the level talent is defined, there is a need for development and continuous improvement of a talent.

On the role that HRM may have to attract and retain talent, the main finding is that HRM has to act in coordination with all the areas of an organization, and never as an isolated department. Additionally, respondents present different strategies and practices for HRM, and the organization, to be successful in attracting and retaining talent, but most significant is that the emphasis in all of those is the need for HRM, and the organization as a whole, to see and consider first the person(s) and not only the talent. This positioning of the respondents is aligned with the perspective of human capital [1], in which the view is that employees are unique and valuable in an organization. However, the respondents recognize that HRM practices are influenced by the institutional and cultural contexts of each society and that it is important for HRM principles, policies and practices to have a vertical and horizontal fit [14, 19].

HRM is consider to have an important role in talent management through the implementation of different practices, such as: recruitment and selection of internal or external persons that can potentiate talent in an organization; continuous training of all workers and creation of good working conditions; and fair and attractive practices of remuneration, incentives and benefits. Additionally, respondents call attention to the problems of HRM to focus only on individuals identified as talented, as advocate by some perspectives of talent management [19, 21, 22], since it can be negative to the organizational culture and discourage the other employees, putting at risk teamwork and collaborative spirit.

As a consequence of the difficulty in defining what talent is, the respondents when asked to identify talents, they did it in three main categories: personality with persistence as the most mentioned talent; knowledge, in which to know how to communicate was the most cited; and group, with teamwork as the talent most quoted. But the respondents also identified individuals and organizations as an example of talent, with a considerable list of names of persons and companies being mentioned. When directly asked to identify a business company that in the opinion of the respondents manage adequately their talents, the list is diversified with names of companies from Portugal, but also from other countries, and different activity sectors. The main criteria to select the companies were appropriate leadership and good human resources practices. Regarding the identification of teams/persons with good capacities in managing talent, the diversity was considerable, from sport teams to singers.

However, when asked about the reasons for an organization to waste/lose its talents, the respondents were very unanimous in mentioning poor organization management and poor human resource management, as a consequence of a devaluation of people considered talents or of the skills that characterize them as talents.

Finally, when questioned about their own talents, the most predominant were the ability and willingness to learn, leadership, ability to work as a team, creativity and persistence, tolerance and willingness to help others, from a rather long list. Although the students that participated in this study are not professionals or have studied talent management, their perception is important and shows the relevance of both practice and theory of IHRM in this particular subject.

The participants of this study will be, most likely, employees of twenty-first century organisations that are facing important challenges in what concerns human capital, and particularly talent management. One key aspect of strategic human resources policy is the maximization of the talent of individual employees as a unique source of competitive advantage for organisations. Therefore, it is important to better understand how organizations effectively manage their talents and also that both organizations and employees/individuals are aware that the context is one of the increasing global competition where talent management plays a central role in organizational and personal success. Both organizations and individuals have to be able to identify, develop and continuously improve their talents, as in the words of the Portuguese poet Fernando Pessoa [42: p. 148]:

> To be great, be whole; don't exaggerate
> Or leave out any part of you,
> Be complete in each thing. Put all you are
> Into the least of your acts,
> So too in each lake, with its lofty life,
> The whole moon shines.
>
> 14-11-1933

References

1. Dries, N. (2013). The psychology of talent management: A review and research agenda. *Human Resource Management Review, 23,* 272–285.
2. Ashton, C., & Morton, L. (2005). Managing talent for competitive advantage. *Strategic HR Review, 4,* 28–31.
3. Farndale, E., Scullion, H., & Sparrow, P. (2010). The role of the corporate HR function in global talent management. *Journal of World Business, 45,* 161–168.

4. Sparrow, P. R., Brewster, C., & Harris, H. (2004). *Globalizing human resource management*. London, UK: Routledge.
5. Kim, K., Park, J.-H., & Prescott, J. E. (2003). The global integration of business functions: A study of multinational businesses in integrated global industries. *Journal of International Business Studies, 34,* 327–344.
6. Frank, F. D., & Taylor, C. R. (2004). Talent management: Trends that will shape the future. *Human Resource Planning, 27,* 33–41.
7. Lewis, R. E., & Heckman, R. J. (2006). Talent management: A critical review. *Human Resource Management Review, 16,* 139–154.
8. Cappelli, P. (2008). *Talent on demand: Managing talent in an age of uncertainty*. Boston, MA: Harvard Business School Press.
9. Collings, D. G., & Mellahi, K. (2009). Strategic talent management: A review and research agenda. *Human Resource Management Review, 19,* 304–313.
10. Tarique, I., & Schuler, R. S. (2010). Global talent management: Literature review, integrative framework, and suggestions for further research. *Journal of World Business, 45,* 122–133.
11. Brewster, C., Sparrow, P., & Harris, H. (2005). Toward a new model of globalizing HRM. *International Journal of Human Resource Management, 16,* 949–970.
12. Scullion, H., & Collings, D. G. (2006). *Global staffing*. London, UK: Routledge.
13. Stahl, G., Bjorkman, I., Farndale, E., Morris, S., Paauwe, J., Stiles, P., et al. (2007). *Global talent management: How leading multinationals build and sustain their talent pipeline*. Faculty & Research Working Paper, INSEAD Working Paper Series.
14. Schuler, R. S., Jackson, S. E., & Tarique, I. (2011). Global talent management and global talent challenges: Strategic opportunities for IHRM. *Journal of World Business, 46,* 506–516.
15. Chambers, E. G., Foulon, M., Handfield-Jones, H., Hanklin, S. M., & Michaels, E. G. (1998). The war for talent. *The McKinsey Quarterly, 3,* 44–57.
16. Michaels, E., Handfield-Jones, H., & Axelrod, B. (2001). *The war for talent*. Boston, MA: Harvard Business School Press.
17. Guthridge, M., Komm, A. B., & Lawson, E. (2008). Making talent management a strategic priority. *The McKinsey Quarterly, 1,* 49–59.
18. Beechler, S., & Woodward, I. C. (2009). The global 'war on talent'. *Journal of International Management, 15,* 273–285.
19. Cooke, F. L., Saini, D. S., & Wang, J. (2014). Talent management in China and India: A comparison of management perceptions and human resource practices. *Journal of World Business, 49,* 225–235.
20. McDonnell, A., Collings, D. G., & Burgess, J. (2012). Guest editors' note: Talent management in the Asia Pacific. *Asia Pacific Journal of Human Resources, 50,* 391–398.
21. Iles, P., Chuai, X., & Preece, D. A. (2010). Talent management and HRM in multinational companies in Beijing: Definitions, differences and drivers. *Journal of World Business, 45,* 179–189.
22. Iles, P., Preece, D., & Chuai, X. (2010). Talent management as a management fashion in HRD: Towards a research agenda. *Human Resource Development International, 13,* 125–145.
23. Cappelli, P. (2009). A supply chain model for talent management. *People and Strategy, 32,* 4–7.
24. Huselid, M. A., Becker, B. E., & Beatty, R. W. (2005). *The workforce scorecard: Managing human capital to execute strategy*. Boston, MA: Harvard Business Press.
25. Björkman, I., Ehrnrooth, M., Höglund, M., Mäkelä, K., Smale, A., & Sumelius, J. (2013). Talent or not? Employee reactions to talent identification. *Human Resource Management, 52,* 195–214.
26. Dries, N., & Pepermans, R. (2008). 'Real' high potential careers: An empirical study into the perspectives of organizations and high potentials. *Personnel Review, 37,* 85–108.

27. Marescaux, E., De Winne, S., & Sels, L. (2013). HR practices and affective organisational commitment: (When) does HR differentiation pay off? *Human Resource Management Journal, 23,* 329–345.
28. Chuai, X., Preece, D., & Iles, P. (2008). Is talent management just "old wine in new bottles"? *The case of multinational companies in Beijing, Management Research News, 31,* 901–911.
29. Huang, J., & Tansley, C. (2012). Sneaking through the minefield of talent management: The notion of rhetorical obfuscation. *International Journal of Human Resource Management, 23,* 3673–3691.
30. Urquhart, C. (2013). *Grounded theory for qualitative research: A practical guide*. London: Sage.
31. Myers, M. D. (2011). *Qualitative research in business and management*. London: Sage.
32. Sparrow, P. R., & Balain, S. (2008). Talent proofing the organization. In C. L. Cooper & R. Burke (Eds.), *The peak performing organization*. London: Routledge.
33. Sloan, E. B., Hazucha, J. F., & Van Katwyk, P. T. (2003). Strategic management of global leadership talent. In W. H. Mobley & P. W. Dorfman (Eds.), *Advances in global leadership* (Vol. 3). New York, NY: JAI.
34. Rioux, S. M., Bernthal, P. R., & Wellins, R. S. (2009). *The globalization of human resource practices. Special report for development dimensions international*. Bridgeville, PA: DDI.
35. Boudreau, J. W., & Ramstad, P. M. (2005). Talentship, talent segmentation, and sustainability: A new HR decision science paradigm for a new strategy definition. *Human Resource Management, 44,* 129–136.
36. Silzer, R., & Dowell, B. E. (2010). *Strategy-driven talent management*. New York: Wiley.
37. Legge, K. (1995). *Human resources management: Rhetorics and realities*. London: The Macmillan Press.
38. Lepak, D. P., & Snell, S. A. (1999). The human resource architecture: Toward a theory of human capital allocation and development. *Academy of Management Review, 24,* 31–48.
39. Creelman, D. (2004). Return on investment in talent management: Measures you can put to work right now. Human Capital Institute research report, Washington, DC.
40. Rawlinson, R., McFarland, W., & Post, L. (2008). A talent for talent. *Strategy + Business, 52,* 21–24.
41. Ribeiro, J. L., & Gomes, D. (2016). Other organizational perspectives on the contribution of human resources management to organizational performance. In C. Machado & J. P. Davim (Eds.), *Organizational management: Policies and practices in a global market*. London: Palgrave Macmillan.
42. Pessoa, F. (1946). Odes de Ricardo Reis. Fernando Pessoa. Notas de João Gaspar Simões e Luiz de Montalvor, printed 1994, Ática, Lisboa.
43. Pascal, C. (2004). Foreword. In Schweyer A (Ed.), *Talent management systems: Best practices in technology solutions for recruitment, retention and workforce planning*. San Francisco, CA: John Wiley & Sons.
44. Duttagupta, R. (2005). *Identifying and managing your assets: Talent management*. London: PricewaterhouseCoopers.
45. Warren, C. (2006). Curtain call: Talent management. *People management (March)*, 24–29.
46. Jerusalim, R. S., & Hausdorf, P. A.. (2007). Managers' justice perceptions of high potential identification practices. *The Journal of Management Development, 26*, 933–950.

Job Market, Generations, and Talents

Sandra Araújo and Sara Oliveira

Abstract We focused our approach on the crosscutting of three variables, job market, generations, and talents. We explore how fast the technological evolution, fueled by the big data and artificial intelligence, is already revolutionizing the law of supply and demand of talent, condemning to failure some tasks or jobs as we know them today and rising others. This technological impulse will imply changes in the management of human capital, so it is necessary to understand the contexts that shaped the present generations in the job market so that the two variables of the equation can be understood. In what it comes to the emergence of talent, we explore the critical skills of a volatile job market and reflect on the expectations that new generations have about it.

1 Introduction

In this chapter written by four hands representative of two different generations, we share a reflection on the future challenges of managing people in an environment of great complexity, especially due to the variables present in the equation: volatile job market, five generations coexisting in the same organizational environment and the need to align talent with the emerging changes.

Without any ambition of technical perfection or to assume incontestable truths, especially because we are human resources professionals who are faced on a daily basis with numerous challenges in terms of managing people in our wide range of customers, the aim of the present work was to provide some reflections on the existing mismatch between what organizations are seeking in their employees so that they can be able to adjust to this change spiral that we are facing and, on the other hand, what these employees expect from the organizations they are working in.

S. Araújo (✉) · S. Oliveira
EDIT VALUE Formação Empresarial, Braga, Portugal
e-mail: sandra.araujo@editvalue.com

C. Machado (ed.), *Competencies and (Global) Talent Management*,
Management and Industrial Engineering, DOI 10.1007/978-3-319-53400-8_5

Perfectly aware of this reality, we are being forced to rethink our role, our skills, and our future actions because it will be required from us the abilities to transform the trends of the external environment and the expectations of internal stakeholders in internal actions. Considering this premise, we will first advance with an approach to the volatility of the job market, followed by the need to transform the human capital in the organization's success key. Such an assumption would be meaningless if we do not know the characteristics of the generations in the job market, once we must respect the past inheritances and their impacts in the future.

It is also very important to know what skills the future will require from the professionals of today and tomorrow, so we will explore the skills that the job market will increasingly value while searching for talent. Finally, we will portray what the future generations most value, once we will have to focus both on the business outcomes and on the human development. To do so, it is very important to analyze the individual and organizational abilities.

2 The Volatility of the Job Market

The job market is changing at a dizzying pace and more dramatically than at any other time in history. Everyday we witness the emergence of new technologies that are leading to significant changes in the way we live and work. We are facing such a ferocious and fast progress that organizations are being pushed to develop the skills of a workforce that does not follow the same pace of the technological development.

Unprecedented advances in terms of artificial intelligence (AI), robotics, Internet, autonomous vehicles, 3D printing, nanotechnology, biotechnology, computing, among others have been redesigning the work processes, blurring boundaries, and creating new opportunities. This moment is dubbed by many authors as the fourth industrial revolution, already underway, which can generate major challenges to the human resources management because it can have significant impacts in what it comes to the relationship between the individual and the work and among people at organizations.

The first industrial revolution brought a wave of innovation by the invention of the steam engine which had impacts on the urban and rural areas, leading to the industrialization of the agriculture, the massification of the urban centers and education. It also made emerge a liberal economy without job regulations, with degrading working conditions and despotic job relations, where the most valuable skills were the physical abilities of each worker.

The second industrial revolution, marked by the development and mass production of the electric light, spread new social models and ways of working. Skilled workers with constant learning needs so they could easily adapt to the machines being used were appreciated, but they developed a mechanical job, physically strenuous and that did not require much thinking. During this period, all organizations' gains were focused in standardized productions where the employees should demonstrate skills such as agility, speed, and a handful of hard skills.

The third industrial revolution, the digital revolution, which we have been witnessing over the past decades, breaks down barriers and makes the world more global by changing the concept of distance. With a simple click everything can change, so the information became more collective and it was necessary to familiarize the professionals to deal with the overload of information and learn how to use it. This event also motivated the creation of new models of organizational communication, both externally and internally, giving increased value to skills such as problem solving, negotiation, flexibility, adaptability to change, and creativity.

The fourth industrial revolution will drag systemic changes in the working structures and, consequently, in the professions as we know them today. With the work processes increasingly robotized, platforms that manage data, smarter systems, and less dependent on operator intervention, some professions are threatened and others will emerge.

It is clear that the need for talent in certain categories of work is being followed by the high instability on the skills to be developed in the future workforce.

According to the report The Future of Jobs, published by the World Economic Forum in [9], five million jobs will be lost until 2020 and one-third of the necessary skills to the job market will change due to the impulses of the fourth industrial revolution.

The use of technology has already changed when and where the work is carried out in almost all sectors of activity giving place to more flexible work practices, such as teleworking.

There are jobs that will quickly become irrelevant and obsolete due to the fractures that the advances in technology, artificial intelligence, nano- and biotechnology, genetics, and other trends in constant acceleration will produce, in the short and medium terms, in the consumption and production patterns and of course, in what it comes to employment. However, new professions will emerge surprisingly quickly and without an urgent action to manage the short-term transitions and build a workforce with adaptable skills to future developments, the governments will have to deal with the rising unemployment and social inequalities and organizations with lack of talent. Actually, this is already a trend recognized within the human resources management that is already facing recruitment difficulties due to the shortage of talent, but according to this report, it will tend to get worse over the next five years. About the concept of talent, it will be explored later.

Despite the fact that the change will not happen in all sectors at the same in the same way, it is certain that the impacts are collateral and they will be felt because the pace of the change is fast and waits for no one, which will force the organizations to be more proactive in what it comes to recycling their acquirements, improving their skills and managing their human capital.

If any certainty exists, is that organizationswill have to focus on talent management and on the reinvention of its workforce, especially in what it comes to the existing set of skills today as many of them will become obsolete tomorrow. The truth is that organizations should not face the fourth industrial revolution with the same bag of tools and the same way of thinking that made them get this far. It will not be enough. As much as they advocate, they will have to open doors to the human capital management that will have to concentrate its forces in the volatile talent management of different generations coexisting in the same organization.

In this equation, the senior talent should not be neglected, as the baby boomers are still present in the job market and it is important to take advantage of their wisdom and experience to build effective processes of knowledge management. Similarly, and although there are already being undertaken some changes in what it comes to education, they will not be sufficient and may still not impact in Generation Z, so it is very important for organizations to do not merely wait for the entrance of this generational group in the job market.

We all have talents, more or less visible, and they must be identified in each of the generations present in the organizational context. In order to build hybrid profiles that are able to adapt and adjust to different situations, it is important to get to know the potential talent and manage the diversity of knowledge, skills, and competencies.

It is also crucial to anticipate strategies and plan actions for this transition, as it is extremely important to know and mobilize generational diversity that necessarily will respond to different stimuli and motivations.

The management of human resources will have increased work on this demand to know the person as a whole. This process begins in the recruitment and extends to the integration, career, and skills management. Getting to know what makes people move, what brings them joy and satisfaction and what compounds their luggage of skills and the competencies they will have to work is the challenge of today and for the next few years. It is also important to keep in mind that the expiration date of skills is becoming increasingly shorter.

To conduct this task so that organizations will be able to capitalize the new opportunities presented by the technological development, it is necessary for the human resources management to take an increasingly strategic role and ensure a place at the table during administration meetings in order to identify the talent gaps and align talent management strategies. If that does not happen, how do organizations believe they will attract, develop, and retain their talent?

This is not about predicting the future, but about structuring the processes that will allow mapping the skills of the different generations and their different expectations, the necessary knowledge to the constant changing environments, understanding which are the current levels of certain skills and which profiles of competencies can and should be improved, so that human resources can add strategic value to the business model.

3 Human Capital: The Key to Success

In the present socioeconomic context, we are increasingly compelled to focus in the short term. This is leading the human resources management to a state of constant change where they need more than just limit their action to the classic processes of recruitment, selection, training performance evaluation, among others. Nowadays, it is mandatory for the human resources management to seek a balance between these activities and the emerging areas of the individualize management of people and, therefore, create genuine human capital.

The changes in the external environment and in the organization's processes design, which are increasingly uncertain and volatile, require the review and constant adaption of the people's knowledge to new roles and challenges and the importance to understand the emergence of new functional identities, much more complex.

Therefore, a new paradigm arises: How can we transform the human capital in the success key of an organization?

It is important to start by understanding each and every word. This will allow us to direct our actions and understand the reason why we nowadays talk about human capital and not only in human resources. The word *resource* transports us to the idea of a path to achieve a certain end. The lexicon has changed, but above all, the context of the organizations has changed and so have people in the organizations. Nowadays, people feel the need to perform a meaningful work in a positive organizational environment, where happiness and respect for their work prevails.

Analyzing the definitions of each word:

- Capital refers to accumulated assets or advantages, likely to produce profits or benefits;
- Human refers to a primate, sociable, which is distinguished from all other animals by the faculty of language and intellectual development.

If people in organizations manifest the possibility to add value and to produce profits by optimizing their intellectual abilities, in what perspective should we see them: as an investment or as a cost?

Similarly to any organization's investment, every person who is a part of the organization must be seen as an investment, a talent that must be accepted, integrated, and prepared to develop its duties. However, attract talents is no longer an easy task. It is necessary to think of a strategy that will enhance the recognition and reputation of the organization in the community.

People working in organizations are the organization brand 24 h a day, 7 days a week. They are the ambassadors of the successes and failures, and therefore, it is necessary to develop a strong and unique organizational culture, based on a system of values, rules of conduct, and management policies that will influence the way people think and act as a whole team, and prepare them for the real transformation that is required.

It is important to create, urgently, favorable conditions for the success of each team that will reinforce the way co-workers communicate among the team and with each other. It is impossible to fight technology, but it is possible to use it in our own advantage to help people stay connected and not lose their interpersonal skills. Bizarre? No!

The challenges in what it comes to human capital are, mainly, to foster a culture aligned with the vision, mission, and strategic goals of the organization in order to identify the necessary and appropriate skills to pursue the previous defined strategy.

In this context of constant changes, it is important to be ensured that the isolated and independent actions of the organization are integrated and sustained through disciplined change processes. It is also relevant that the internal capacity to deal with change is equal or even greater than the external pace of change. Acting on human capital implies helping to make those changes at an institutional level, changing standards at the initiative level and making thing happen at an individual level, enabling personal change. This implies working over the organizational culture, making clear the importance of change in order to overcome resistance and involve everyone in this process. It is necessary that organizations, through people, become more agile, flexible, responsible, and able to implement changes in order to create sustainable value.

Well, to build a culture of change with which people in the organization can relate and that each team member believes in will allow:

- To guide the attitudes of employees, making them ambassadors of the organization's brand;
- To transform the organization's talents in behaviors toward the implementation of the organization's values in the actions they take on a daily basis;
- To accentuate the identity of each member of the organization and the sense of belonging, distinguishing the company from others by the diversity of talent that will result in innovation.

Therefore, the management of human capital has to be strategic and align the policies and practices with the talents of the organization.

Transforming the human capital into a talent source implies understanding the networks of cooperation and confidence, that is, the social capital as well as the psychological forces such as self-confidence, emotions, expectations, optimism, hope, authenticity, resilience, that is the psychological capital.

This impels the human resources management to explore, listen, and understand the behavior of each of the team members and the influence of the leadership and decision-making model in the team structure and, particularly, in each person. And we are not talking about a sanctified leadership [6], in which leaders are seen as heroes who are not allowed to make mistakes and from whom perfection is required at all times. We are talking about courageous leaders (and many of them are), who are human, sometimes endowed with some imperfection and weaknesses inherent to every human being, but that can generate empathy and bubbling emotions. While sharing positive emotions, the talents and individual skills become more visible and that will promote the effort in achieving the outlined results.

Hoping that the comparison does not result in misleading interpretations, it is known that even the modern dairies are adapting to this change context and are now equipped with beds for their animals. And these are not just ordinary beds. They meet technical standards so that the animals will not suffer any kind of harm that may affect the quality of the milk produced. In that same environment, the animals listen to classical music so that their stress levels decrease and will not affect their productivity and contribute to their happiness. It is true that the ultimate purpose is the profit, but isn't it the purpose of any organization?

The challenges are huge! Add to the organization's DNA an energetic spirit centered on the development of the talent from several generations that need to be more adjustable to the contexts of change, leverage this generational talent, transform all of this human capital in the organization's success key and turn all of this into a strategy that will produce positive results, it is not an easy task. However, this is the challenge of today and for the times ahead, regardless of how it is approaching!

For this reason, contextualize expectations and behavior patterns of each generation in the labor market will help trace paths between supply and demand. On the one hand, by identifying the needs that different talents seek at work and, on the other hand, the skills that the job market will most value while searching for that talent.

4 Managing Generational Diversity in Organizations

Very soon, and for the first time in history, five generations will be working side by side. Nowadays people are living longer and, therefore, working until later. This event led us to this new situation. However, by putting members of different generations working together we are both presented with an opportunity and a challenge: the opportunity to involve a group of people with unique experiences and skills and the challenge of dealing with the generational differences that distinguish them. We have been trying to define what factors drive each generation and what they value the most at a professional level and what we know by now is that every generation that enters the labor market adds to it a unique set of motivations and gains and, consequently, the differences between generations can affect the way organizations recruit and form teams, deal with change, motivate, manage people, and encourage productivity.

At organizations are currently converging the following generations [7]: veterans (born before the year 1945), baby boomers (born between 1946 and 1964), GenXers (born between 1965 and 1979), Millennials (also known as

Generation Y, born between 1980 and 2000) and Post-Millennials (also known as Generation Z, born after 2000). It is important to refer that due to the young age of the Post-Millennials, the studies about this generation are still inchoate.

These five generations,[1] formed in completely different social, economic, and cultural contexts and that, according to those contexts, have developed different skills learned to, consequently, value disparate skills. In this regard, it is noteworthy that the historical events have influenced in a profound way the values of every generation. These events have the power to refer to people emotions and memories and are able to shape the way people feel the importance of family, career and even money. And it is enough to think about the events that marked these generations to realize that the needs of the job market also suffered very significant changes. For example, to an administrative that is a veteran, the use of the computer was not a reality until many years after the beginning of his career while currently it is basic computer skill for any administrative born in Generation Y. As we can see, several professions have undergone profound changes with the technological developments—some have fallen into disuse (typists and archivists) and others are now setting their position (social network manager, for example).

Understand and appreciate the factors that influenced the contexts that have shaped each generation can be a great help for organizations since the development of recruitment procedures to the implementation of the organization's talent motivation strategies. Similarly, it can still be a relevant contribution in the recognition of their gains and how to maximize it, urge their success, and involve people in the organization.

The constant search for competitive advantage, driven by a changing context, encourages organizations to deal with people management as seriously as they do with technological developments. Along with the changes in the social values and global economy, it is also important to reflect and constantly reformulate the needs of human resources. The evolution and growth of professionals will largely depend on the effective performance of the managers, aiming a healthy working environment and aligned with the organizational goals. This is the current challenge for the upcoming years so that the organizations keep their workforces motivated, in order to put the best skills of each generation to the organization's service and, consequently, obtain a better performance from the entire team.

We will briefly examine the Generations baby boomers, X, Y and Z. It is important to note that the generation veteran is already retired or about to, so it will not be focused on the present analysis. Similarly, and regardless the generation they operate in, people are different and their needs are also diverse, so we cannot assume that the members of each generation are equal. We can just assume that, due to the context in which they were born and grew up, it is possible that they share similar perceptions about the world.

[1]It is important to refer that the time periods between the generations can suffer slight variations.

4.1 Generation Baby Boomers (1946–1964)

The Generation baby boomers include the people born between 1946 and 1964. After the end of World War II, the USA experienced an exponential increase in their birth rate, a fact that was associated with the soldiers return to their homes and that lead to the name of the generation. Baby boomers are identified as the inventors of the peace and love era, mainly because of their aversion to armed conflicts. They preferred music, arts, and all other forms of culture and were the responsible agents for major changes, starting with the debate of the women's role, breaking down political barriers. This was the generation in which people primarily left their homes to live alone. They were protestors and that catalyzed a series of changes in society.

The boomers are the veteran's children and underwent a period of economic growth and positive changes, but they have also witnessed the Vietnam War, the launch of the Sputnik satellite and gave life to the hippie movement. They were political and social protesters by the peace movement, lived the rock and roll explosion and the growth of television [2].

Born between 1946 and 1964, the baby boomers are now beginning to retired. Due to the historical, cultural, economic, and social context that accompanied them, they share similar characteristics. In particular, they tend to be people who value the hierarchical positions and traditions, the durability of their interpersonal relationships at work, permeated by mutual exchanges, and reciprocal values. For this generation, changing jobs must be properly addressed. Respect for the older people and structured processes (especially within organizations) are also valued aspects. In what it comes to their career, they are loyal workers and respect authority, while also seek to challenge the system in which they operate.

Professionals from the Generation baby boomers look for stability in their careers and a steady job in which they can develop their careers and stay until retirement. They are not particularly motivated by new challenges or the idea of being promoted to leaders. For this generation, the most important is to be recognized for the experience and not for their creativity or capacity to innovate. However, this does not mean that they are accommodated and have no prospects for their future.

Baby boomers represent a different time, when there was less competition in the job market, the economy was booming and the options for professionals were not as many as today, so having a stable career was the intended. Paralleling, we can say that baby boomers wanted to grow, but they did not care if they had to spend several years performing the same tasks, as long as they had stability. In what concerns to Generation Y, the situation is slightly different. The Generation Y is not afraid to leave a good job if it feels that other organizations may be even better or that it will be able to conquer or achieve goals that drive them to action. Baby boomers are known for their work ethic and in what it comes to the balance between the personal and professional life, is safe to say they are mainly focused on work.

Boomers tend to embrace change and, such as the veterans, have a commitment to work that includes loyalty to the employer. However, and despite this similarity to the veterans, boomers prefer to be seen as equal.

It is possible to conclude that baby boomers were very important for the future generations. Thanks to them, it was easier for Generation X to adapt to their reality and set their own desires and goals.

4.2 Generation X (1965–1979)

Generation X was born between 1965 and 1979. Although there is no consensus on the time period that effectively covers, people born since the early 1960s and until the late 1970s are included in this generation. Those born in this period represent the pop culture of the 1970s and are often dubbed as *latchkey kids* because they have often been left alone at home, once both parents worked, which explains their independent and adaptable approach to work. Sometimes referred to as the *lost generation*, this was the first generation of children exposed to situations of divorce and to the concept of children day care. During childhood, Generation X has become the most unprotected in modern history. This generation was marked by the assassination of Martin Luther King and Robert Kennedy, the Watergate[2] political scandal and the first man landing on the moon.

This generation has learned to be independent since a very young age and turned that value into a remarkable brand as it advanced in the world of work. When Generation X first began to emerge in force in the labor market, the economic decline of 1980 took place. During this period, in which the members of Generation X began to leave the universities and finish their degrees, between 1975 and 1995, the reduction of the employment rate emerged and, consequently, organizations reduced the number of employees and many of the *latchkey* children watched their parents become unemployed. This event triggered a change of values of these young people in what it comes to loyalty and job security. The competition for jobs was shaking. For the first time in history, it was not expected that this generation would be able to reproduce the lifestyle of their predecessors.

Generation X is arguably more qualified than the previous generation. Due to their education and increasing maturity, they are also more conscientious and pragmatic about starting a family. This generation has an entrepreneurial spirit, a *do-it-yourself* attitude and, in contrast to previous generations, embraces change in the workplace. The members of this generation are career-oriented, but place a strong emphasis on family time and strive for a balance between personal and professional life. They enjoy their autonomy and freedom—work to live rather than live to work, which is not always easy for the baby boomers to understand.

[2]The Watergate was a political scandal in the 1970s that occurred in the USA and culminated in the resignation of the USA President at the time, Richard Nixon.

A flexible workplace is important for the member of Generation X, as well as a constructive feedback. Generation X and baby boomers have different work ethics—Generation X moves for diversity, challenges, responsibilities, and creativity while baby boomers have a preference for stricter approaches.

Generation X has suffered a strong influence of the globalized world and the dedication given by their parents to their professional careers. The members of this generation entered the labor market approximately between 1985 and 2000, and they carry as core values the balance between personal and professional life, the importance of family, and the satisfaction and quality at work.

Summing up, Generation X prefers less formality and less conventional hierarchies, placing merit above status and is less loyal to organizations, compared to baby boomers. The members of this generation know how to work with computers and give a strong value to money and benefits, while they seek for balance between work and personal life.

4.3 Generation Y (1980–2000)

This generation is constituted by people born in the early 1980s and until the late 1990s and it also known as Millennials. It is expected that by 2020, it will constitute half of the active population. This generation was born exposed to diseases such as AIDS, witnessed terrorism as a global phenomenon, and grew up playing with electronic equipment. However, proactivity, ambition, and optimism are characteristics associated with Generation Y.

Generation Y has grown up with the Internet facilities, entertained by electronic games, and the technology is incorporated in their lives. The members of this generation believe in constant change, and their focus is on a short term. Plus, they are not fans of pre-established rules and the frequent exchange of jobs is seen as natural. They are not people used to face authoritarian environments and do not like to wait—they are used to have everything at a distance of a click.

The children that constituted this generation had a schedule full of activities, such as music or language lessons, and sports, so the time to play without having to demonstrate productivity was scarce, especially due to the need that families felt to prepare them promptly to the future.

Millennials developed the systemic thinking, very encouraged by the virtual world, being that competence relevant to the workplace. However, they need to develop their emotional intelligence. Also noteworthy are the behaviors of openness to diversity such as gender, race, beliefs, or religious differences.

Generation Y tends to highlight the fundamental goals of life for which its members are willing to fight, such as a comfortable life and liberty. It is known as the generation technology for having been born and grown up immersed in technology advances—the members of Generation Y entered the job market approximately around 2000.

Technological advances and changes in society have had a great influence on Generation Y and when compared with previous generations it tends to have a more tolerant behavior, more confident, and more biased to value education. Furthermore, it is composed by people with higher skills and integrated into the virtual technologies, as well as the various types of technology and, in those situations, can have a greater ease and skill that their predecessors had.

Once the Millennials have grown in an environment of technological advances and enhance learning that carries the expectation of continuing this process of continuous learning within the organization. In short, it is important for them to feel that their professional and intellectual growth is aligned with their needs. In short, it is important for them to feel that their professional and intellectual growth is aligned with their needs.

Because it is a generation used to speed, it is important for them to receive opinions, suggestions, and worth improvement aspects about their performance at work on frequent basis. On the other hand, the balance between the delegation and the freedom of action is also an important strategy to keep the Millennials involved. It is essential to provide freedom of action to the professionals of this generation, so that they can put into practice the power of innovation and creativity that characterizes them, after determining a goal to be achieved and a deadline for a particular activity.

Millennials are people with creative ability and a taste for challenges. They need to feel motivated in the workplace and will not hesitate to look for another opportunity if they do not feel recognized. The typical member of Generation Y is productive and highly goal-oriented, seeks personal growth, a career with meaning and a mentor/leader to encourage and facilitate its personal development.

4.4 Generation Z (After 2000)

Formed by young people who were born after 2000, this generation is characterized by the letter Z, which was originated from the term *zapping*.[3] This generation has a different perspective on how to face the world. The members of Generation Z do not appreciate conventional vertical hierarchies. They are young people who talk to the supervisor in the same way they would talk to a subordinate.

Despite not knowing much about the performance of Generation Z in the job market, we know the context they are growing in and it is expected for them to be more confident and assertive in what it comes to their goals, especially compared with Generation Y, plus they will have in their possession more information about their potential employers.

[3]The term used to describe the act of changing channels fast using the remote control without paying much attention to a program in specific.

Similarly to Generation Y, the members of Generation Z want their work to make a difference and to have a meaning, but they are also more concerned about job security than the previous generation. This happens because they have experienced a financial crisis and, therefore, tend to appreciate more the idea of developing their skills and build their career in just one organization.

They do not know the world without Internet and do not differentiate the online and off-line life. They are dynamic, critical, demanding, and know what they want and do not appreciate inflexible schedules.

The entry of this new workforce in the job market will impact due to the peculiarity of their characteristics and will require organizations to adapt and apply new practices to attract and retain these professionals. This generation will not submit to working conditions that will not satisfy its needs, once the members of this generation know what they want. Unlike Generation X, which accepted the rules, the members of Generation Z are not afraid to question and make good points.

These digital natives are less driven by money than Generation Y and have more entrepreneurial ambitions. The proactivity toward digital media also leads many of them to wish to have their own startup. Unlike Millennials, digital natives do not have in mind the concept of work hard, play hard.

This latest generation was born with the Internet, smartphones, online games, and a whole range of gadgets—digital natives were born at the speed of knowledge and speed is their normal pace.

The members of Generation Z are more conscientious about money and spend cautiously—the previous generation grew up in a time of strong economy and the current Generation Z grew up in complex and volatile times. Those born in this millennium do not want to give up their free time and do not consider that work hard and stay in the office hours after the end of their shift is rewarding.

5 Valued Skills When Searching for Talent

If the technology is changing, if the job market needs are changing, if the new generations have different needs, it will be in everyone's interest if competencies and skills are also adaptable.

The relationship between individual and technology has been contributing to the development of closed worlds where social interaction is made through the online network where everything is allowed to say or do. Feedback is less constructive and the bonds of trust between people are attenuated between reality and virtuality. From our work experience, we have found, although empirically, that soft skills such as communication, teamwork, resilience, and the management of emotions are tools increasingly required to integrate the labor market and deal with the everyday problems that all organizations have.

The new generations are coming to the job market hungry for knowledge and new challenges and adventures. They bring in their luggage dreams, expectations, and goals that they want to achieve in the distance of a click. They paint in their minds a screen where the work environment springs happiness, but with abstract representations of what is an organization and how it moves inside and out. Increasingly confined to the online communication where an e-mail serves as an element proof of what has been told. They appreciate to have the phones in their ears, because that is the only way they can focus and because others do not share the same musical tastes. But they want the same thing that previous generations wanted: attention, commitment, sharing, teamwork, motivation, and to demonstrate that it is in the relationship with others that they will get the strength to do different and reflect their true talent.

Deming [3], associate professor at Harvard University, has argued that the soft skills will be crucial to the new job market where people will have to move between different roles and projects.

Even though the future workplace may be seen as increasingly virtual, at home, in the car, on the phone, but at the distance of a click, it will continue to be required **teamwork skills** to solve problems that arise at a breakneck speed and that will provide clarity at the organizational environment, involving all but still masked by the so-called *cloud*.

From empirical data collected in organizational contexts, relating to organizational climate evaluations, which allowed us to develop a benchmark analysis among the organizations that we follow annually, it seems reasonable to say that the indicators show a declining trend in what it comes to the collaborative relationships between employees (the average rate of the communication index is positioned in an intermediate zone registering 2.9 on a 4-point scale and the cooperation index and teamwork registered 3 points). However, these are indexes that have been decreasing since we first started to develop this work within our clients' universe. Indeed, it is a complaint that has become common among the affected teams.

Aware of the importance of communication and collaboration to face the ongoing changes, it is important to disseminate this expertise of teamwork that includes in itself collaboration skills, communication, confidence, and commitment.

In concrete terms, the challenge is to find working models that integrate the interactions in something common, such as the launch of a new product/service and the planning of common tasks.

In this regard, Bohm [1] states that "… *we do not know how to live together in a changing world. We only know how to live based on the truths from the past, which inevitably results in the present in a group that seeks to impose their truths to another group.*"

Once the world is spinning faster than ever, it is difficult to be sure of the functions of tomorrow and which technical skills will be more appreciated, but there is a global awareness that the critical skills to the job market in 2020 will focus on soft skills, which then will feed the learning capacity of the hard skills. According to the World Economic Forum, skills such as the ability to solve complex problems, critical thinking, creativity, coordination with others, and

Table 1 *Source* Future of Jobs, World Economic Forum

Top skills in 2020
1. Complex problem solving
2. Critical thinking
3. Creativity
4. People management
5. Coordinating with others
6. Emotional intelligence
7. Judgement and decision making
8. Service orientation
9. Negotiation
10. Cognitive flexibility

emotional intelligence will be the most valued in the professionals of tomorrow because it will allow people to move into the unknown and constant innovation (Table 1).

Creativity will become one of the most valued skills once it will facilitate the emergence of new ideas, new products, new technologies, and new ways of working, adapted to more informal and flexible contexts.

Likewise, **emotional intelligence** is one of the main skills, once the widespread idea is that emotions can make thoughts more intelligent. But to develop it, it will be necessary to understand ones emotions and the impact of these in them and in others. It has to do with the ability of emotional intelligence to facilitate the thought, once the cognitive system, problem solving ability, reasoning, and decision making are influenced by emotions. Emotions are part of people and we should seek to integrate them to the rational to obtain gains.

Empathy will take a strong commanding voice. The technology can erode the sense of empathy, the ability to analyze the world wearing someone else's shoes. But it will be important to engage the teamwork processes, operational changes, and feed the stronger interpersonal relationships to ensure the commitment to the task and to people.

On the stage of uncertainty, where the functions may change frequently, **resilience** will be one of the most requested skills to all of those who operate in the job market.

The challenges that businesses face and will face are much more complex than in the past, ruling the ambiguity of situations and contexts, which requires change as each year goes by, so **adaptability** will be a key competence, which will enable organizations to face big changes and find comfort in increasingly dubious contexts regarding future actions.

With the accessibility of information, where sometimes misinformation reigns, individuals should be able to sift through a large volume of data the most relevant information and take decisions based on these, so **critical thinking** plays a key role.

Another critical competence is **credibility** because it will impact at a trust relationship level. The professionals of today and tomorrow must build their credibility by fulfilling its promises at the accorded time, leading them to establish positive personal relationships and build professional relationships based on clear, consistent, and integrated communication.

Aware that many of these skills are not being focused in the education system, it is important to realize that the investment in them will have to be made by the individual wishing to enter the job market or by organizations absorbing new talent.

One certainty we have in common with many other professionals in the field of human resources, managers, and business managers: It is increasingly difficult to recruit the right person to the right place because the right person does not respond to the right place requirements. And we are not talking about the technical knowledge requirements, these ones may be there, but about personal skills that have high impact on the job performance. For this reason, the focus becomes the individual so that later, and starting from the individuality of each one of us, we are able to build the web of relationships among all that will favor the growth of the organization.

6 Retain Talent, Pleasure Generations

Over the years, there has been an increasing debate about the challenges that the new generations bring as a new workforce and we are now beginning to better understand how the organizations can adapt to the demands of these workers. The knowledge and understanding of the context from which they emerged as well as their needs will be an indicator and an aid to better align the organizational practices with the goals of those who cooperate in it.

The fact is that with an age difference of almost 50 years between the older workers and the youngest, there is a wide range of perspectives, needs, and attitudes that converge in the same workplace. Although generational diversity in the workforce promotes a wider range of talent, recognizing and understanding generational differences can help develop strategies to attract and retain talent in order to achieve competitive advantage in the market.

According to the results presented by Hays Portugal on their "Job Market Guide [4, 5]," 7 out of 10 professionals in assets in Portugal considered the possibility to change job in 2015. Even thought these are not data that we can use as guidelines for all over the world, there is a unanimous opinion that the job change is a constant in this changing world. This desire for change has somehow to do with the inability of organizations to retain their talents, which has gotten worse with the entry of Generation Z in the job market.

Another important aspect is to understand what we meant by "talent." There are several perspectives on the concept of talent, but there is a common thread, as the authors agree that talents are individuals with skills, abilities, attitudes, and differentiated experiences from each other. Ulrich [8] argues that talent depends on three coexistent components in the individual: competence, commitment, and contribution. Skills are the abilities and values necessary for each person to perform their tasks in the organization. The commitment comes from building a value proposition, to ensure that employees give value to the organization and engage in its duties by being aware of how they are contributing to increase the competitiveness of the company and boost its results. The contribution relates to the satisfaction of the employee and occurs when he feels that his personal needs are being met by participating in the organization.

Consequently, it is necessary for the human resources management and the organizations to have the capacity to develop and implement tools that will allow the commitment of each person at the same time contributing to their personal development, once the retention or loyalty to the organization is influenced by the benefits that the organization offers and by the reconciliation with the goals of each person.

Retaining means keeping the talent in the organization so that they can perform in the best way possible their tasks. But how do we retain different generations with expectations and forms of personal satisfaction so diverse?

First, it is important to think about what organizations are doing today to become more attractive, how are they developing their brand, the so-called employer branding (Illustration 1).

Every organization has to communicate effectively what it does and what is different from the competition. But it is important that between the packaging and the product it contains, there is symmetry. And we are not talking about marketing actions, but about what the marketing in line with the human resources can do to attract new talent through the dissemination of its best practices.

To do this, you must first build and develop human resources practices (integration, socialization, career management, etc.) aligned with the people and for the people in the organization and that today require more creativity and transparency. What is happening within the organization must be communicated, and it must be consistent with what people say and feel about the organization in order to do not disappoint expectations and develop a reverse effect.

Illustration 1 Differentiating elements of the brand

The mission and values should be well absorbed by every member of the organization and mirrored in the daily behavior and relationships between different stakeholders.

The organization, as a brand, must carry a benefit, so it is necessary that the wage policy and benefits are attractive. And we are not only talking about the economic benefits, but about the emotional wage, the salary component that is not mirrored in wage receipts and that it is related to the emotions that are awaken with prestige, with social responsibility, and with the organizational reputation.

Attract talent involves betting on the organizational environment, creating positive environments that enable facing times of tension. These aspects are felt through people, their expressions, their smiles, and it contributes to such the internal communication, collaboration and network sharing, the identification with the task, the physical conditions of the workplace, and many others we could list here.

It should also be highlighted the role of leadership and how the leader inspires the entire structure at its most different levels (Illustration 2).

Despite the fact that the wage component has an impact on the attraction and retention of talent, it is certain that what the new generations such as the Millennials or Generation Z seek is the opportunity to continually develop their knowledge and enhance their skills and learn. It is important to focus on developing a training strategy focused on the development of their personal skills, already here identified as strategies for the future, as well as their organizational skills that foster new discoveries and the assimilation of new areas.

Similarly, they seek challenges, the possibility to develop a variety of tasks that enable them to strengthen their core competencies in order to continue to develop a culture of improvement.

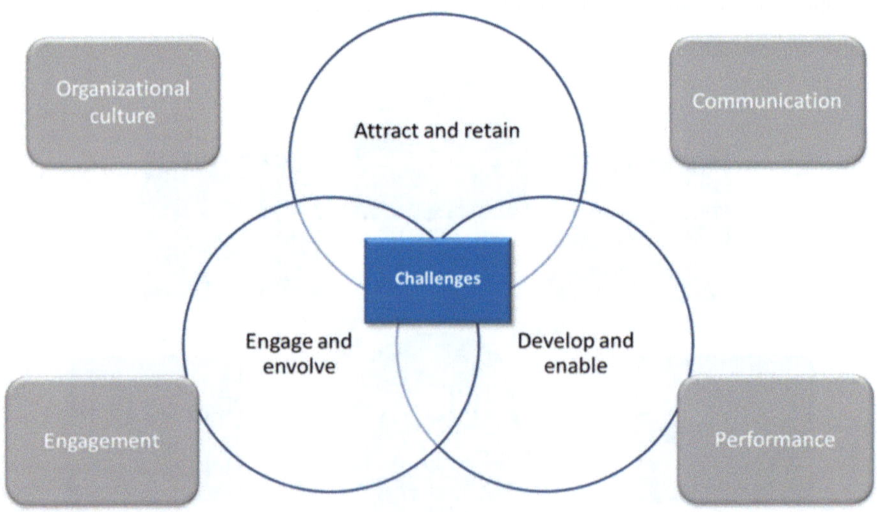

Illustration 2 Challenges to retain new generations

The recognition will always be the imperative word, so it is important that it is given by words, gestures, and attitudes. It is also important to realize that work and family reconciliation is also a valued aspect for the Generations X, Y, and more will be by the Generation Z, so it is increasingly necessary to think of conciliatory practices between personal and professional life.

In the same study of Hays Portugal, it was asked what were the most valued features in a potential employer: A good work environment is highlighted (77%), followed by the wage offer (76%), career plan (70%), organizational culture (60%), and the quality of the projects (52%).

Recognize the work that everyone performs for them to feel that their contribution means something to the organization involves developing daily briefings on projects and work in progress, networking meetings, happy hours, team buildings, celebrate victories together and develop social responsibility practices. Such actions will impact not only on the employee performance level but also on the emotional connection that this will create with the organization, generating impact, at the same time, on the culture and external communication, once organizations are also what people say about them.

Effectively there are no universal recipes, because we are talking about people. The most important is that while searching for talent we identify diverse strategies that are aligned with every member of the organization, once the satisfaction of most will transpose positive emotions that are contagious to the workplace and that will help in the dichotomy of satisfaction and retention of true.

7 Conclusion

In the alignment between the job market and the expectations of the new generations, it is important to take into account that the way to attract and retain talent is mainly done through the development of a strategic human capital management centered on the development of each individual's social and psychological capital as well as skills such as teamwork, emotional intelligence, ability to solve complex problems, empathy, critical thinking, and other soft skills. This process of construction is already imposed today, but it will be even more challenging as new generations integrate the job market so that equilibrium can be found.

References

1. Bohm, D. (1996). *On dialogue*. New York: Routledge.
2. Crampton, S. M., & Hodge, J. W. (2009). Generation Y: Unchartered territory. *Journal of Business & Economics Research, 9,* 1–5.
3. Deming, D. (2016). *The growing importance of social skills in the labor market*. Harvard University and NBER.

4. Hays Portugal. (2015). *Geração Y: um novo paradigm laboral – um estudo sobre as motivações, preferêmncias e tendências da Geração Y no Mercado de trabalho em Portugal.*
5. Hays Portugal. (2015). *Guia do Mercado Laboral.*
6. Rego, A. C., & Cunha, M. P. (2016). Pecados da Santa Liderança. *RH Magazine a Gestão de Pessoas em Revista, 104*, 28–32.
7. Tapscott, D. (2008). *Grown up digital* (1st ed.). Boston: McGraw-Hill Education.
8. Ulrich, D. (2014). The future targets or outcomes of HR work: Individuals, organizations and leadership. *Human resource development international* (pp. 1–9).
9. World Economic Forum. (2016). *The future of jobs: Employment, skills and workforce strategy for the fourth industrial revolution.*

Global Talent Management: Reality or Utopia? A Special Glance Through a Portuguese Multinational Organization

Joana Ribeiro and Carolina Machado

Abstract In this global world that we live in, global talent management is one of the big challenges faced by the human resource management areas around the world. It constitutes a privileged and innovated way of aligning human resource management and business strategy. The main purpose of this study was to identify and analyze the talent management System presented by a Portuguese multinational organization. In addition, there were two other propositions with significant relevance for the case: the analyses of a conceptual framework about human resources roles in global talent management and the impact of contextual factors; and the analyses of the principles of effective global talent management. The theoretical construct was supported by a qualitative approach. Data collection approaches for this qualitative research involved data observation and individual semistructured interviews. All global talent management strategies, policies, frameworks, and practices adopted by the multinational company have been carefully examined in order to pursuit the goals of the research. It was concluded that global talent management is now a reality for this Portuguese company and the company's global talent management system can be used as a model, as far as it concerns to the talent strategy and philosophy, influenced by the different roles played by the human resources area and the observation of the principles of effective global talent management. However, it is absolutely necessary to note that the success of each talent management system depends on each particular situation and on the organizational context as well as on the ability to innovate and on the creativity of the implemented policies and practices.

J. Ribeiro · C. Machado (✉)
School of Economics and Management, University of Minho, Braga, Portugal
e-mail: carolina@eeg.uminho.pt

C. Machado (ed.), *Competencies and (Global) Talent Management*,
Management and Industrial Engineering, DOI 10.1007/978-3-319-53400-8_6

1 Introduction

> There is something that is much more scarce, something finer far, something rarer than ability. It is the ability to recognize ability.
>
> Hubbard [1, p. 163]

Nowadays, one of the biggest challenges faced by organizations all around the world is the fact that human resource management (HRM) trends are directly connected to globalization and, consequently, to the intensification of the complexity of business activity.

The success of international expansion is based on the way the key people are managed. To attract and retain key talent, companies worldwide must recognize the importance of international mobility, as well as the adoption of talent management programs, and make efforts to create a talent mobility strategy [2]. According to this author, the talent pool is not local anymore, instead is becoming more and more global.

2 Literature Review

2.1 Global Talent Management

Human resource management is now facing new challenges, highlighted by the increase in global employment competition and also by the demand for talent. According to the SHRM [3], the competition for the best skills has been particularly emphasized by the rapidly emerging economies, such as China, India, and Brazil.

2.1.1 Definition and Evolutionary Perspective

Global talent management was widely accepted by human resource practitioners and consulting firms worldwide. Building on their excellent work, academics recently started to examine the talent management phenomena more closely in last few years [4].

Although there is a growing consensus that global talent management is an emerging area, there is no consensus regarding the exact definition or boundaries of it [5].

According to a literature review made by Tarique and Schuler [5], global talent management "is about a systematically utilizing IHRM activities to attract, develop, and retain individuals with high levels of human capital (e.g., competency, personality, motivation) consist with the strategic directions of the multinational enterprise in a dynamic, highly competitive, and global environment" [5, p. 124].

The international mobility of key people, with the intention of developing some individual or organizational business areas, skills or needs, is a crucial part of a talent management strategy.

Managing a global workforce is a big challenge, because it goes beyond what concerns to the geographical and cultural boundaries. One of the main research topics in this field is the lack of talent [5]. According to the Manpower Talent Shortage Survey [6], "Despite the continuing caution exercised by many companies amid going economic uncertainty, a substantial portion of employers in the U.S. and worldwide identify a lack of available skilled talent and a continuing drag on business performance." According to the same source, the most difficult positions to fill in the Americas include technicians, sales representatives, and skilled trade workers, in the Asia-Pacific include sales representatives, technicians and laborers, in and Europe, Middle East, and Africa include skilled trade workers, technicians, and engineers.

Despite how talent management is defined, this is an emerging field which has been creating new roles and jobs. A recent study by Heidrick and Struggles [7] notes that we are beginning to see the steady emergence of a new discipline of strategic talent management, led by a Head of Talent or similarly titled role. "The cumulative impact of global demographic trends, combined with ongoing economic uncertainty and aggravated by a critical skills shortage, creates a powerful talent triple whammy facing business. As a response, companies worldwide are bringing talent, mostly leadership talent, to the top of the agenda and are assigning responsibility for aligning business and talent imperatives" [7, p. 2].

However, Guthridge et al.[8] identified many barriers to the use of HR policies and practices of global talent management initiatives such as the following:

- The fact that senior managers do not spend enough time on Talent Management maybe because they think there are more pressing things to be concerned with;
- Organizational structures that inhibit collaboration and the sharing of resources across boundaries, which include based regions, products, or functions;
- Middle and front-line managers who are not sufficiently involved in or responsible for employees' careers because they see these kind of activities as less important than managing the business;
- Managers feeling uncomfortable on evaluating and acknowledging performance differences among employees;
- Managers at all levels who are not sufficiently involved in the formulation of the talent management strategy and therefore have a limited sense of ownership and understanding of global talent management actions.
- HR departments that lack the right competencies to address the global talent management challenges effectively.

Sculion and Collings [9] emphasized that HR policies and practices must be mutually supportive and internally consistent. These also need to fit the specific firm characteristics, such as management leadership, values, vision, culture, size, and type of industry. More specifically, it would be expected from them in their HR policies and practices selection to:

- Identify the specific global talent challenges that are confronting them;
- Evaluate their rigor and sophistication levels;
- Determine firm characteristics that are likely to be more supportive of certain HR policies and practices;
- Continually monitor the drivers and shapers of the firm's global talent challenges, facilitating the changes needed to implement them.

Therefore, global talent challenges emerge in the context of a dynamic environment. There are significant HR strategic business issues that ensure just the right amount of the right talent and motivation, at the right place, at the right price, during all economic and financials ups and downs, balancing the workforce with the needs of the firm in the short and in the long term [10]. In this context, significant Talent Challenges have emerged, specifically those associated with the need to reduce and remove talent in order to lower the costs of operations; locate and relocate the operations around the world; and obtain equally competence talent anywhere in the world at lower wages.

According to the mentioned authors, the major forces and shapers of the global talent challenges are as follows:

(1) Globalization. Globalization is a concept that people use when referring to different phenomena. For this issue, it has been giving particular relevance to expanse of world trade, intensified competition among firms, the potential to reach more customers around the world, and the set of people worldwide who now compose a global labor market.

(2) Demographic Changes. In the next years, with the exception of India, societies will be "greyer" than Japan, currently the country with the oldest population, and businesses worldwide will face a "greying" workforce. At this point, few companies offer lifelong opportunities to keep skills current. On the other hand, there a big issue with the Generation Y' expectations: The lack of developing opportunities is the major reason given by Generation Y employees for leaving a company. Corporate Universities are turning their attention to attracting and developing the members of different generations [11].

While the populations of many developed countries are aging and shrinking in size, the populations of developing and emerging economies are expanding and getting younger [12]. There are significant variations in demographics characteristics by age and by region all around the world that multinational organizations need to consider in locating and relocating their operations internationally [13].

(3) Demand for workers with competencies and motivation. New jobs are still being created that require higher levels of technical competencies. For the existing jobs, there is a growing need for employees more adapted to change and under new conditions that require the development of additional

competencies [14]. The need for highly motivated employees is likely to remain strong as well. Highly motivated or highly engaged employees, with high levels of productivity, are able to contribute for more to the firm than those who are less motivated [15].

(4) Supply of workers with competencies and motivation. In developed economies, such as Western Europe, North America, and Japan, there is an expected shortage of managerial competencies, especially as the economy recovers.

In this context, Tarique and Schuler [5] suggested that global talent management, as any emerging field, requires much exploration to improve the clarity of definitions, frameworks, and models. According to them, there are several topical areas for future research to better understand the global talent management including:

- What it means to be a bridge field;
- Using more theoretical frameworks;
- Identifying more specific differences;
- Moving beyond descriptive statistics;
- Developing the notion of "systems" in global talent management systems;
- Implementing global talent management in different country contexts; and
- Developing exist strategies for talent.

2.1.2 Principles of Effective Global Talent Management

Stahl et al. [16] suggest six principles of effective global talent management. According to these authors following talent management, best practices can only take you so far. Top-performing companies subscribe to a set of principles that are consistent with their strategy and culture. One of the biggest challenges faced by organizations worldwide is building and sustaining a strong talent pipeline. The authors recognized that adopting a set of principles rather than best practices challenges current thinking. Best practices are only the best in the context for which they were designed for. On the other hand, principles have broad applications. The authors found out that successful companies adhere to six key principles: alignment with strategy; internal consistency; cultural embeddedness; management involvement; balance of global and local needs; and employer branding through differentiation.

1. **Alignment with strategy**. Corporate strategy is the natural starting point for thinking about talent management. Given the company's strategy, what kind of talent do we need? This requires a significant overall of existing performance management systems, investment in line management capability, and overall changes to the mind-set of line managers and employees.

2. **Internal consistency**. Implementing isolated practices may not work and can actually be counterproductive. This principle refers to the way the organization's talent management practices fit with each other. Consistency is crucial.
3. **Cultural embeddedness**. Many successful companies consider their corporate culture as a source of sustainable competitive advantage. They make efforts to integrate their core values and business principles into talent management processes such as hiring methods, leadership development activities, performance management systems, and compensation and benefits programs. Rather than selecting employees for attitude and cultural fit, a more common approach to promoting the organization's core values and behavioral standards is through secondary socialization and training.
4. **Management involvement**. Talent management processes require managers' involvement, at all levels, including the CEO. Senior leaders, and line managers need to be actively involved in the talent management process and make recruitment, succession planning, leadership development, and retention of key employees their top priorities.
5. **Balance of global and local needs**. For multinational companies, operating in different countries, cultures, and institutional environments, talent management seems complicated. Companies need to figure out how to respond to local demands while maintaining a coherent HR strategy and management approach. A company′s decision about how much local control to allow depends partly on the type of industry; for instance, consumer products need to be more attuned to the local market than pharmaceuticals or software. Companies that find a balance between global standardization and integration and local implementation have the best of both worlds. They can align their talent management practices with both local and global needs, resulting in a deep, diverse talent pool.
6. **Employer branding through differentiation**. In order to attract employees with the right skills and attitudes, companies need to find innovative and creative ways of differentiate themselves from their competitors.

On the basis of the above, it was concluded that companies cannot simply mimic top performers, because best practices are only the best when they are applied in a given context. Companies need to adapt talent management practices to their own strategy and circumstances and align them closely with their leadership and talent philosophy, while at the same time finding ways of differentiating themselves from the competitors [16].

2.1.3 Human Resources Roles in Global Talent Management and Contextual Factors

This study was based on a research developed by Sparrow et al. [17] that expands the knowledge of the process of talent management on a global scale which explores a framework of human resource roles in global talent management designed by Farndale et al. [18].

Sparrow et al. [17] focus on the role of the HR function in facilitating the successful management of key talent across the organizations based on data from two large multinational firms based in the UK and the USA. A particular point of interest is how the HR roles may be changing in global talent management based on the global financial services crisis.

Therefore, Talent Management has become more critical than ever. According to Somaya and Williamson [19], some argue that given the large number of lay-offs as a result of the crisis, there is no longer the “war for talent,” popularized by McKinsey. However, organizations continue to seek the top, justifying the significant investment made in the most highly competent people [17].

Farndale et al. [18] developed a conceptual framework that extends beyond the specific global talent management strategies. They described four roles which may be adopted to facilitate global talent management in multinational organizations:

(a) ***Champion of processes***: Developing and monitoring global talent management practices and policies, tools, and strategies; ensuring that these are implemented across the organizations; monitoring GTM processes; and improving coordination of tools, processes, and techniques across functions.
(b) ***Guardian of culture***: Ensuring a mobility culture across the organization; incorporating values in organizations strategies and activities to support global mobility of individuals.
(c) ***Manager of internal receptivity***: Encouraging the inflow and outflow of key talent across business entities; encouraging receiving units to manage diversity, careers, integration, and work–life balance.
(d) ***Network leadership and intelligence***: Developing appropriate networks inside and outside the organization to support the GTM process; being aware of developments in the internal and external labor market; mobilizing appropriate talent both internally and through external provider and a sense of timing and context.

Sparrow et al.’s [17] biggest challenge was to test this framework empirically, to see how the roles, if present, are actually enacted, and to explore at what extent different organizational contexts impact on the framework: business model, talent philosophy, and international financial crisis.

According to the same authors, talent can be divided into two perspectives, focusing either on the subject, individuals with career management needs, or on the object, knowledge, or competencies that the organization needs to manage. Part of the distinction depends on whether people are seen as having innate abilities or whether everyone has strengths which can be developed to reach the top talent status. Talent management then differentiates between taking a generic or differential approach to the management of potential. Should an organization aim to develop everyone, adopting a very inclusive approach, allowing everyone to get the chance to rise to the top; or should development opportunities be offered exclusively to the best? Both approaches constitute talent management systems, but each requires a different implementation strategy.

According to Stahl et al. [16], an exclusive approach, centralizes the rewordings and the attention on the top talent. Otherwise, it assigns less recognition, rewordings, and development needs to the other employees. On the inclusive approach, organizations' main concern is to try to develop and reward all the employees, on an equal way.

The two Talent Management philosophies can live together. Many companies combine them. Regarding specific talent pools (senior executives; specialized technicians; young talents), there could be different development strategies. A hybrid approach helps on the differentiation and overcomes the big controversy on the intrinsic value of some groups or functions.

Within the global talent management field, one factor differentiates it from "domestic" talent management which is global mobility or expatriation—the movement of key people to overseas locations of the organization [17]. Managing global mobility becomes a key part of the human resources strategy for global talent management. However, there is no single business capable for operating internationally, and there can be several variations in the roles of the HR function in different types of international organizations [17]. The distinction is made between organizations with more centralized operations compared to those which are more decentralized, with local operations able to decide their own approach to global talent management.

According to Farndale et al. [18], the centralized model requires high global integration, while the decentralized model requires local responsiveness. The general trend is toward increasing centralization in order to have some control on the development of the future leadership of the organizations.

Sparrow et al. [17] examined whether and how adjustments to the global integration and local responsiveness balance are being made in the financial and professional service firms. They also explored the extent to which the enactment of each of the HR roles is likely to depend on the contextual and strategic factors.

The first proposition relates to the need to build a core of competence able to transfer capability across multiple countries. In order to achieve this, the four roles are expected to be present.

The second proposition builds on the first. Although the four roles may be present, the authors anticipate the variation in the prevalence of these roles based on the context of the firms: inclusive or exclusive approach; centralized or decentralized business model; and the specific characteristics of the financial and professional service firm contexts.

The final proposition suggested that these four roles will be dynamic and the 2008 finance crisis is likely to have changed the global talent management Agenda and the supply of talent and reduced available resources for implementing global mobility strategies.

According to Nijveld [4], these four roles should be in good balance. The alignment between those roles leads to organizational performance. But the alignment between those roles is not the only alignment there should be. Internal alignment is the key for a positive relation between GTM and organizational performance.

Human resources area is a crucial competitive advantage source on the design of business strategy settled on three main issues: the belief on the potential of human development; the intention to value people on their workplace; and the reinforcement of leadership on the development of organizational culture [20].

According to Stahl et al. [16, p. 2], the competitive advantage in talent management does not come from implementing best practices, rather it comes from internal consistency of all the elements that constitute the talent management system, in other words, the way the talent management practices fit with each other.

Best practices as recruitment, staffing and succession planning, training and developing or retention management are not the key to competitive advantage, but they must align closely with the various elements of TM system, such as business strategy, leadership philosophy, and value system of the organization.

Schuler et al. [10] believe that the organizational culture in the form of business needs and strategies is also a main characteristic on the design of the talent management strategy. According to a literature review made by these authors, there are also many external contingencies that influence global talent management: economic trends, competitiveness, labor market conditions, and national culture.

Talent management globalization requires a specific approach that considers additional international pressures. In the current economic climate, it may no longer be appropriate to talk about a "war for talent." This expression was replaced by "demand for talent" as a result of two big challenges: local competitiveness and the new forms of mobility related to the emergent markets. More people available on the labor market does not necessarily mean that employers are able to find the level of skilled professionals they are seeking. The demand for talent remains high and it stills remaining a lack of talent, especially on the high technology field in the emergent markets. There is also a high talent competitiveness beyond expatriates and local workers [21, 22].

Farndale et al. [18] believe that the quickest step to internationalization and globalization involves a strategic role of the human resources area. An effective talent management system allows the multinational companies to reach a global competitive advantage [10].

3 Research Methodology

The theoretical construct was supported by a qualitative approach. The purpose of the research is simultaneously descriptive and analytical.

The purpose is descriptive at the first part of the research, where the intention was to obtain and describe the main characteristics of relevant questions, namely through the identification and description of the company's global talent management strategies.

The adopted methodological processes were selected in order to contribute to achieve the proposed empirical knowledge. In order to respond to the proposed purposes, the methodology was fundamentally qualitative and it is related to the phenomenological paradigm. According to the qualitative characteristics of the required information, the case study is the core methodological instrument of this research. Regarding to the process, this is a qualitative research, based on the perceptions examination. On the other hand, regarding to the purpose, the research is analytical, intending to analyze, explain, and understand different phenomena and finding cause–effect relationships. The case study consists on an extensive review of the phenomenon on a Portuguese multinational company. The reasons behind this choice are as follows: the company's intern and international market projection and collaboration on University of Minho projects.

Data collection approaches for this qualitative research involved data observation and individual semistructured interviews, including two thirty-minute nondirective exploratory interviews, to the Talent and Leadership Development Coordinator and to the Head of Talent Management and Development, with the intention of present the case study. These were followed by a conference call, directed to a Talent and Leadership Development Specialist, in order to get more information, and followed by an hour semistructured interview to the Head of Talent Management, with the purpose of testing the followed model and to better understand the way how the company attracts, selects, promotes, and retains high potential workers. The semistructured interview script was based on the script adopted by Sparrow et al. [17]. The questions focused on global talent management practices, policies, and strategies; business context and talent strategy; and on the core functions of the Talent Management and Leadership Development area. This was followed by data qualitative analyses.

4 Analysis and Discussion of Results

Beyond the Literature Review and after the presentation of the Research Methodology, below follows this case study explanation, analysis, and discussion of results.

The empirical research took place on a prestigious Portuguese multinational company.

4.1 The Company

- **Presentation**

The case study's company is a Portuguese multinational founded in 1959. It is a retail company (food retail, specialized retail and retail properties) with two core partnerships: shopping centers and telecoms. In order to assure the confidentiality and anonymity rights of the company, we decided to call it "Y".

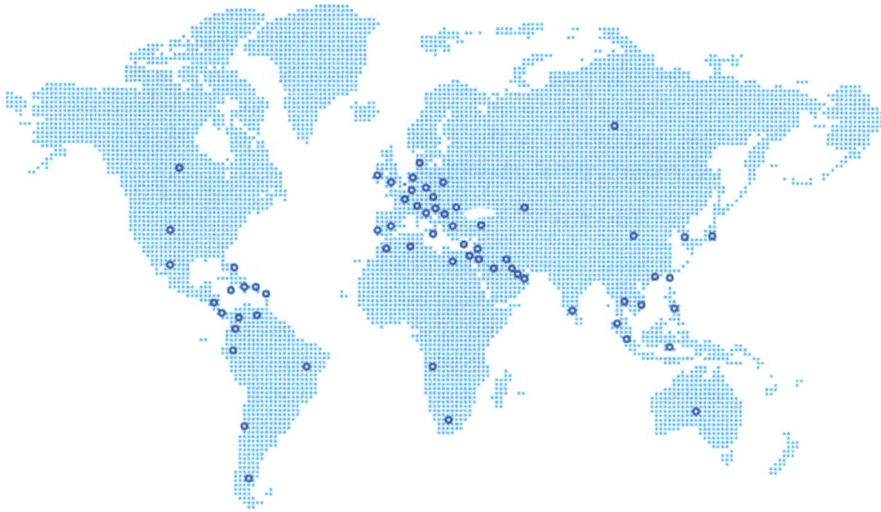

Fig. 1 The Y world. *Source* Management Report'14

The company's corporate strategy is based upon the creation of value through three strategic pillars: international expansion, diversifying the investment style, and leveraging the asset base in Portugal [23]. Therefore, internationalization is the foremost strategic priority for future growth, and it focuses on core businesses and adjacent businesses areas.

At the end of 2014, the company was active in 67 countries (Fig. 1), including operations, third-party services, representative offices, franchising agreements, and partnerships.

According to the Chairman's Letter presented on the Management Report [24, p. 19]: "Success comes from valuing our team. This is a living organization and we are committed to ensuring that every single colleague feels that they are important to our success."

4.2 Human Resources Management: Integrated Talent Management

At the end of 2014, the company had 40,947 employees, featuring a young team (52% young people under the age of 35), and characterized by diverse profiles and gender (62% women).

In 2014, the company had 38,726 employees in the retail area. Of these, 67% were women and 8% were international collaborators, namely from Spain, Brazil, China, and Turkey [24].

The company's sustainability strategy has three axes of performance: better purpose, better planet, and better people. The biggest better people commitment is "the integrated talent management," promoting well-being and investing in the development of employees' skills and abilities, continuously enriching the company's culture [24]. According to the Sustainability Report [23], the integrated talent management commitments are as follows:

- **Attracting Talent Effectively**

The company has been given particular importance to the creation of direct contact opportunities between the business world and students and recent graduates. Through the promotion of different initiatives, the company supports personal and professional development of young people and identifies young talent. The company played an active role in bridging the gap between young people and the labor market by:

- Participation in job fairs at schools and universities;
- Promotion of workshops and providing support for academic and research work;
- Promotion of study visits to the company (central structures and operations);
- Undergraduate, Master, and MBA students can participate in mentoring programs;
- Partnership in the "Bué d´Escolhas" program, promoting initiatives focusing on providing skills and development of children and youngsters at-risk;
- Participation in the partnership regime of the "Alliance for Youth," which aims to contribute to the fight against youth unemployment in the Portuguese and European context.

- **Career Management**

The company's careers' model plays a role in supporting the management of the talent pipeline, ensuring alignment between priorities and professional development expectations. Therefore, this reflects two management approaches: One focused on the planning needs of human resources and the other more directed toward the employee as an actor of their own professional and personal development.

The company values the diversity of profiles and pathways of its people and supports personal and professional development between areas, directing the attention toward reflecting on this model and the internal mobility process that allows the employees to acquire cross-cutting and multipurpose skills.

- **Training and Development**

The company has demonstrated the ability to generate innovative and entrepreneurial leaders internally which has proven to be significant to the rapid development and diversification of the business and occupying leadership positions in various segments of activity.

This fact is due to investing in the best management practices and development of talent, but more importantly by the introduction of this concern in the daily lives of the leaders.

For the company, "being a leader means developing talent and recognizing that the achievement of goals depends on the performance and commitment of extraordinary teams, who have to constantly overcome the challenge of efficiency and continuous improvement" [23].

With emphasis being placed on the following:

- Management and leadership academy;
- Training center—continuous improvement;
- Forums for knowledge-sharing and consulting groups;
- Innovation and creativity management and the sharing of in-house knowledge;
- Retail school; and
- Mentoring program.

- **Performance Evaluation**:

Based on the principles of meritocracy, pluralism, and participation, the company's performance manage model encompasses all the employees, in all the businesses and locations. Every year, the company invites all the employees to reflect on their achievements and aspirations, ensuring a performance management interview with each team member. During this interview, an opportunity is given to discuss the results achieved in the year under review, sharing expectations and ambitions, the identification of development opportunities, and the definition of goals and action plans.

Within the framework, the companies have had place since 2012 a talent management tool, a nine-box grid that despite encompassing all the employees was mainly designed for the most senior employees in the organization and for the employees identified with the most growth potential.

4.3 Talent Management and Leadership Development: The "Talent Nine-Box Grid"

It was found that the human resources department is a big differentiating asset for this company.

The human resources administrator reports directly to the CEO, which is equivalent to a business CEO. He has sat on Executive Board, and, at the same time, he leads the HR Advisory Board, chaired by the CEO and Chairman, and responsible for the definition and approval of all the HR policies, the majority of which are related to talent management policies and contribute to the alignment with the global management strategy. As a result, the communication of talent management issues is facilitated, as though the sharing of ideas and concerns.

The **Talent Management and Leadership Development** area led by the Head of Talent Management and Leadership is responsible for the development and implementation of strategic talent management policies through the different HR departments, with a special focus on Top Executives. This area supports on the conceptualization and operationalization of employee's personal development models, according to the company's strategic guidelines and the best practices, in order to maintain and improve the talent pipeline of the organization.

However, it is important to highlight that the company, under the talent management and leadership development area, positions all the employees on a "Talent nine-box grid" which forms a very organized talent management System. The "Talent nine-box grid" is a matrix tool that is used to evaluate a company's talent pool based on two factors, performance and potential.

The "Talent nine-box grid" is part of the Performance Management [25]. The Performance Management System's target public includes all the employees, from team members to senior executives. The system applies equally to all the employees that have had a minimum of six months of exercise of functions. The evaluation stands on an annual basis with a formal follow-up in the middle of the year.

Therefore, in order to get a careful analysis of the company's talent management system, we precede to a detailed analysis of the Performance Management System which crosses performance and potential.

The performance evaluation system [25] intends to adjust cross-cultural skills to the particular needs of each business. Thereby, it includes two different kinds of components:

(1) Evaluation components (individual goals; soft skills; technical skills).
(2) Information components (self-assessment; performance review; career development; development needs; reviewer's comments; employee's comments).

The employee's global performance grade comes from the qualitative result of partial results obtained from the components Individual goals, soft skills, and technical skills expressed on the next performance scale [25]:

(1) Unsatisfactory;
(2) Needs improvement;
(3) Effective;
(4) Very good; and
(5) Outstanding.

The informative components are from major importance to the company because they assure data qualitative collection that complements the qualitative components which enables the connection between all the HR policies, particularly Career paths, and Training and Development.

In this regard, and with the exception of self-assessment, all the informative components are mandatory to this process. Self-assessment is recommended and preliminary to the performance evaluation interview.

On the informative component, career development takes place the discussion of the employee interests on global mobility, as well as the fit to that kind of career.

Therefore was created a mobility assessment which is the result of the evaluation of a crucial range of soft skills and technical skills to this issue, such as the following [25]:

(1) Commitment to organizational principles;
(2) Striving for excellence;
(3) Intercultural understanding;
(4) Communication skills;
(5) Ability to network effectively; and
(6) Foreign language skills.

The evaluation of performance system, as a connecting tool through all the HR processes, intends to identify the employees' training and development real needs and the functions performed requirements. Thus, all the skills grades under 3 (reference level) will be listed as development needs.

Another crucial parameter to the **Talent nine-box grid** is potential assessment. To a company, a high potential employee is somebody who can rise to the top and can succeed in senior and critical positions.

Potential is the result of aspiration, engagement, and skills, which increase success probability on a senior position [25].

The potential assessment methodology adopted by the company is mainly explained by the following [25]:

- On the aspiration and engagement fields, there are three levels: 1—little; 2—medium; and 3—high;
- On the skills field, it takes place an automatic analysis to the employee skills gaps, through the difference between the level 4 on the skills evaluation (exceeds) and the employees average grade on the soft skills and technical skills evaluation: gap > 1: employee scores 1 on skills; gap > 0.6 $\leq$ 1: employee scores 2 on skills; and gap $\leq$ 0.6: employee scores 3 on skills.

The potential assessment result is automatically presented by the calculation of the partial average scores of each potential field [25].

- If the average result is between [1–2.5], the employee is on level 1;
- If the average result is between [2.5–2.8], the employee is on level 2; and
- If the average result is between [2.8–3], the employee is on level 3.

On level 1, it is not expected the rise of the employee to major responsibility and complexity functions within two or three years.

On level 2, if it takes employee development, it will be expected the rise of the employee to major responsibility and complexity functions within two or three years.

On level 3, it is expected within two or three years, and the rise of the employee to major responsibility and complexity functions.

Thereby and according to the scores of the employees on the Performance Evaluation and Potential Assessment, they are framed on the talent nine-box grid (performance vs. potential) [25] which follows below (Fig. 2).

This is a talent nine-box grid, 3 × 3, because the two lower scores (unsatisfactory and needs improvement) and the two higher scores (very good and outstanding) are agglutinated.

The employee's framework on the talent nine-box grid is a valuable input, not only to the definition of priority policies but also as to the supporting tool to all the HR policies and practices.

The talent nine-box grid includes different segments to which match a specific color and designation (Fig. 2).

Therefore, at the end of the performance evaluation and the potential assessment, after the talent nine-box grids were properly completed, take place the Talent Forum Reviews at the performance management cycle. Through a global vision is discussed and adjusted the organization's talent, based on a range of relevant indicators and analysis.

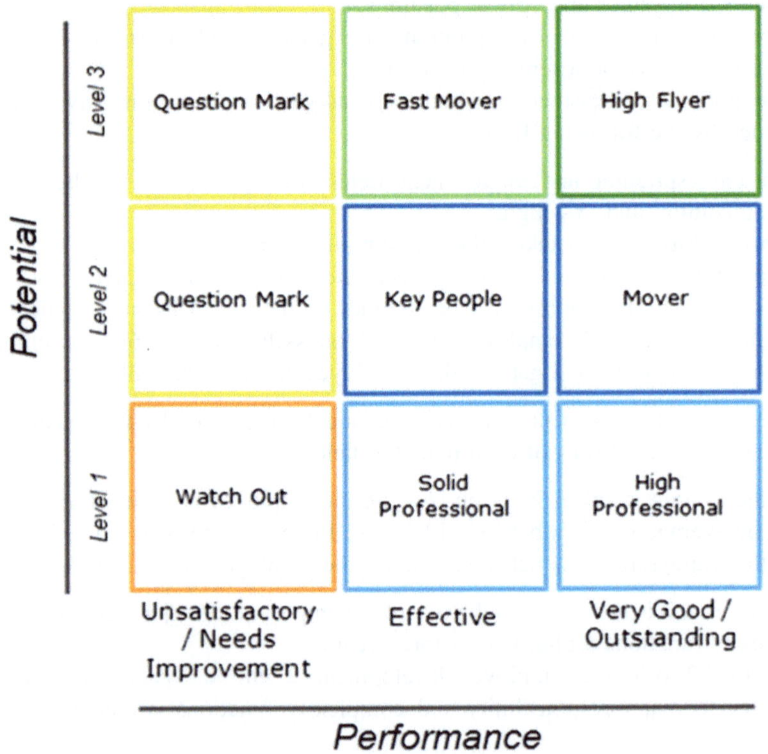

Fig. 2 Talent nine-box grid. *Source* Performance management system manual [25]

The Talent Forum Reviews' main goal is to identify and discuss the top talent management, with special focus on the next talent nine-box grid segments [25]:

- High flyers (employees with excellent performance and high potential who can make commitments on risky challenges in new areas or on their own) and fast movers (employees that achieve the expected performance levels and with high potential, who are able to take risks and can have different professional paths).
- Key people (employees that achieve the expected performance levels and reveal medium potential levels; they should strengthen their current function) and movers (employees with excellent performance and medium potential levels who can be eligible to horizontal mobility in the short run with the same complexity level).
- Questions Marks (employees who reveal medium or high potential, but who did not have the opportunity to show it yet; they reveal low-performance levels that could be related to some kind of functional mismatch or because they are a new admission).

Thereby, all the macro-action and development plans are designed for the employees. This process follows and incorporates the organization's strategic planning cycle.

In this context, it is important to highlight the Management and Leadership Academy's role on the leadership development, at the extent of training and development and career management fields. The Management and Leadership Academy's main goals are as follows: contributing to the dissemination of organizational culture; creating moments of sharing experience; building networks among employees; and promoting internal mobility, in order to increase the synergies between different business areas.

4.4 The Company's Main Talent Management Challenges

The big obstacle assumed by the Head of Talent to the talent management strategy's implementation is the huge dimension of the company. According to her, "plural mindsets impact on information systems and constrain the implementation of inclusive, concerted and unique models, however, that is easy to overcome."

Additionally, some challenges were diagnosed on the implementation of the talent management strategy related to get a good balance between the following:

(1) Internal talent versus external talent;
(2) Individual work versus teamwork; and
(3) Young employees versus older employees: generation gap and diversity management.

All of these challenges are key issues for the company. There is a clear strategic focus on in-house development, but sometimes, the company feels the need of external recruitment. The company manages to balance internal talent and external talent as well as assures the mobility between different areas or businesses. Many times, the need of external recruitment comes from the special requirements of new business areas.

With regard to the individual work vs teamwork, the Head of Talent was emphatic "Success comes from the accomplishment of teams." For the company, the teamwork is a core issue to the achievements of goals. However, good teams always have individual talents.

Change demographics and generation gap are part of international agenda. The company has recently made a diagnosis and is working on the necessary adjustments in order to get generation equality.

In Portugal as well in all developed nations, we have an aging population and, consequently, an increase in the retirement age. The company has been assessing the developments in this situation and has been designing an action plan. This action plan focuses on the alignment, optimization, and adjustment of all talent methodologies, ensuring maximum benefit of different generations.

4.5 Principles Examination to an Effective Global Talent Management

The main aim of this point is to verify the existence, in the adopted strategy by the organization, of the six principles defended by Stahl et al. [16] to the global talent management efficacy.

1. **Alignment with the strategy**. We have observed that talent management is a key part of the business strategy. Human resource administrator reports directly to the CEO, being equivalent to a business CEO, and is a member of the board of directors, having access in all executive commissions. On the other hand, there exists a forum called Human Resource Consulting Group, headed by the CEO with the leadership of the human resource administrator, who defended and approved all HRM policies, whose majority reports to talent management policies, making easier the alignment with the global management strategy. We have observed, this way, the existence of a very close line top/down and bottom/up, facilitating the communication of talent management themes and the concerns share.

In the light of the foregoing, we have observed an effective alignment of the talent strategy with the global strategy. About the talent nature, the organization adopts a hybrid approach. Inclusive in the sense that everyone could be a potential talent, thus positioning all collaborators in the Talent Matrix, which crosses performance with potential. But at the same time exclusive considering the particular attention given to three specific talent segments such as, senior executives, high potentials, and persons considered critical and that deserve a more customized follow-up. Depending on their positioning in the Talent Matrix, there are persons that are the target of more customized processes and others of more global processes. The global process, included in the *Dragging our talent* program, corresponds to the matrix extremes, being monitored by their superiors. The particular three segments earlier referred are directly monitored by the *talent management and leadership development* area, being the target of a strategy designed to the growth and to leverage more critical themes.

2. **Internal consistency**. According to this principle, implement practices in an isolated way can be counterproductive. All talent management practices need to be adjusted to each other—the consistency is essential. The organization Performance Management process covers all the organization, about 40,000 persons, with close rates in the order of 95%. It has as main aim to manage and develop talent, in the attempt to conciliate personal aspirations, development and career opportunities, mobility, and business development, in the win/win sense.
3. **Cultural embeddedness**. Organizational culture is considered as a source of competitive advantage to Y organization. The organization develop efforts to integrate its values and principles in the talent management processes, such as, recruitment processes ("Contact Program," "Call for Solutions," "Call for Summer"), leadership developmentactivities ("Management and Leadership Academy," "Knowledge-Sharing Forums and Consulting Groups," "Mentoring Program"), performance management system ("Upward Feedback," "Tracking on Talent"), compensation and benefits programs. The organization looks to promote its values and competences through socialization and training.
4. **Management Involvement**. Talents Management process highlights managers commitment, in their diversified levels, including CEO. Leaders are highly involved in the Talents Management process, since talents identification to the definition of succession plans for them, as is possible to observe from the Talents Matrix and Talents Revision Forums analysis presented earlier. These place their persons in the Talent Matrix and receive feedback from the teams. Indeed, everybody is directly involved, regardless the leadership level.
5. **Balance between global and local needs**. The organization faces two main needs: answer to local demands and maintain a coherent HRM strategy. The multinational organization in study looks to instill its global standards, recognizing simultaneously the need to do the necessary local adjustments, providing the subsidiaries with the autonomy needed to maintain local specificities.
6. **Employer branding through differentiation**. The organization uses the employer brand as a marketing strategy. Therefore, it has found creative and innovative ways that have enabled distinguished themselves from their competitors and become attractive to potential talents, namely through the adoption of creative and aggressive strategies in talent attraction (specific individual profiles; business students; diversified talent pool, including more and more women); the talent philosophy included in the global strategy; and the belief that organizational culture is a source of competitive advantage.

4.6 HRM Roles in Global Talent Management Examination and the Influence of the Organizational Context

This present study looked to analyze the different roles that can be performed by HR managers as global talent management facilitators through the balance sheet analysis between global integration adjustments and local responsibility needed to an organization that acts globally.

In this regard, and from the interviews analyses and the diverse documents made available by the organization, we have easily saw that the organization philosophy transfers capacity by its collaborators and the different countries where it acts and presupposes the existence of the presented four roles:

- Monitor the implementation of relevant policies and practices ("*We have a Corporate area, within Human Resources, Talent Management and Leadership Development, which positions all persons in a talent matrix and supervises directly three segments of talents that deserve a special attention through very customized and variable processes. The remaining are subject of a more global process supervised by line managers, which focus, always, the development*"—Head of Talent)
- Disseminate the organizational culture ("*We are daily motivated by our history, culture and values wealth. In this sense we dispose what we call of 'our way of doing", a document that summarizes our culture and values and that is based in four crucial pillars: Our values, how do we work, or teams, our managers. We try hard day after day in order to be able to contribute to become our persons distinctive and unmistakable, in all places where we operate, being Talent Management processes prepared to adjust to all international scenario*"—[23, p. 22])
- Establish networks and guarantee that all parties are sensible to global mobility needs ("*We develop in a regular basis transversal forums to the diverse business areas, with the aim of share knowledge and promote the adoption of best practices. We also have four consulting groups, with periodical meetings designed to share and coordinate information. In addition, we count with two commissions used as a platform to knowledge and experiences share*"—[23, p. 28]).

Considering the presence of the four roles, a new preposition was therefore established. In other words, we wanted to know whether it will be possible to foresee the prevalence of each of them, based in the company organizational context, i.e., according to the adopted talent management approach (inclusive, exclusive or hybrid), the talent philosophy or the world economic crises context.

In the multinational organization, the role ***Champion of Processes*** assumes a great relevance. "*HRM, through the Talent Management and Leadership Development, assumes the main role in the Talent Management System architecture, namely in the Talent Matrix design and analysis, previously analyzed, and in* *the follow-up of all involved in the Performance Management Cycle, that covers all the organization—Staff Corporate and stores*" (Head of Talent).

In the Y organization, the ***Guardian of Culture*** role is important, in the sense that it assumes to be critical to manage and conciliate the organizational culture role, which is highly evident and strong in the organization, with the expatriation process and the acquisition of local talents and the needed local adjustments.

It was found that in most situations of the international expansion, the organization resorts to local workers. In those situations when the organization resorts to expatriation, the expatriates are carefully prepared to the international mission. In the talent management point of view and according to the organization Head of Talent, "*the organization considers absolutely indispensable to have an international mindset where the language care (English language is adopted as the official language of the Group) and the respect and attention toward certain cultures are crucial*". Sometimes is possible to observe the need for local adjustments "*in what concerns Spanish people, who have less familiarity with English, there was an attempt in adopt the Spanish language in business*" (Head of Talent).

Human resource managers of the Y organization also assume, clearly, the role of manager of local receptivity as they show flexibility and the necessary complexity in career management, showing adjustment to the new approaches of the international assignments, namely with the existence of more and more self-initiated expatriates.

"*The communication of Talents Management themes is always facilitated as well as the concerns share*" (Head of Talent). Talents Management assumes a crucial role, paying HRM attention to the change of processes, mentalities, and practices, compared to the workers international experiences. This role is well visible in the Talent Matrix and more specifically in the Talent Revision Forums. The organization is concerned in developing future leaders and promoting a talent management culture, which is considered a long-term investment.

Repatriation is also considered crucial: Repatriation success is an important tool to encourage others to international assignments.

Intelligent networks and leadership: HRM reveals an extensive knowledge and applicability of the last tendencies and developments in internal and external labor markets as well as high sensitivity to what is happening simultaneously at the local and global levels. Social and organizational networks assume a crucial role to internal receptivity. According to Y Organization Head of Talent, networks and relations established in the organization can be presented in four levels:

(1) Communication among collaborators;
(2) Networks related with specific roles that allow leaders to identify international individuals and assignments;
(3) Networks with the respective HR Departments, encouraging regular contacts to the discussion of global mobility subjects;
(4) Networks among expatriates that allow the exchange of experiences and concerns, creating a social and professional support.

HR area reveals a concern in managing structural social capital (configuration, density and network function/relations) and, simultaneously, social cognitive capital (shared objectives and culture, such as language, codes, among others).

The final preposition suggests, as it had happen in the study that supports this present one, that roles performed by HR managers' enablers of global talent management are dynamic, product of **the organizational context influence**. In what concerns the context of the global economic crisis, contrary to what happened in the study developed by Farndale et al. [18], in the present study we have concluded that despite the crisis started in 2008 has affected global markets and changed the way to search talents, it did not require a resource reduction in the implementation of the global mobility strategy adopted by the organization. In the multinational study, the great influence is exercised by the organization talent philosophy inherent to the global strategy that is people orientated.

Talent management system of Y organization and the talent philosophy that underlies it presuppose the existence of a hybrid management approach. If, on the one hand, all workers are integrated in a Talent Matrix—inclusive approach, on the other hand the organization focuses its performance in top potentials and senior executives, having to them a customized and differentiated approach—exclusive approach.

The organization further considers that there is no single business model able to operate internationally. The type of management developed by the organization at the subsidiaries level is decentralized, with local approaches to the Talents Management. The organization adopts a decentralized model, providing subsidiaries with some local responsibilities. However, the organization considers essential to have a good balance between global integration and local responsibility. If global integration allows the organization to transfer its management philosophy in the countries where operates, local responsibility allows to provide the subsidiaries with great autonomy and the possibility to incorporate local specificities and culture, having respect by the environment where they operate.

4.7 Final Considerations and Recommendations

After analyzing all the global talent management system of the organization, its philosophy and talent strategy and respective Talent Matrix; evaluating HRM roles enablers of the Talents Management; and verifying the existence of the principles to the global talent management efficacy, some final considerations and possible suggestions of continued improvement were established, as followed in the next paragraphs.

Therefore, the multinational organization should continue its efforts so that its global and local organizational structures, at the products and functions level, allow the collaboration and resources share cross borders. Looking to ensure a continued improvement in its Talent Management system, middle managers need to be increasingly involved in the global talent management process (which is, in this process, unanimously the less strong point in any organization), investing in its training and competences development at the leadership level, performance appraisal, and teams management, recurring also to a less classic and more innovator technics, such as coaching, mentoring, and emotional intelligence development.

In its HRM policies and practices selection, the organization should remain attentive to the great Talent Management challenges (reduce or remove talent; locate or relocate operations in the world; and obtain similar talent in any part of the planet); evaluate, always, the rigor and sophistication level revealed by each policy and practice; seek to identify, always, its characteristics that can best support the implemented policies and practices; and constantly monitor the drivers that contribute to the global talent great challenges (talent scarcity; demographic changes; change of attitude toward job positions; cultural differences among countries), thereby facilitating the implementation of the required amendments. Therefore, the organization should remain attentive to the drivers' impact in Talents attraction, development, and retention. At the talent attraction level, the emphasis goes to the talent planning, the employer brand construction and the talent aggressive search, with the implementation of strategies increasingly creative and aggressive, recruiting and developing foreign people to work in its home country and attracting a diversified pool of talents, with the inclusion of more women.

About talents development, the efforts that have been pursued by the organization in leadership global competences development are considered quite assertive and should be combined with succession and substitution plans.

Talents retention presents a huge challenge to this and any other organization. "*Talents retention depends mainly from our capacity in develop a work environment that promotes personal and professional development*" [24]. It is suggested that the organization continues to use the leadership programs, which are an important tool in talent retention. At the same time, it is of huge relevance to focus in a sensitive issue of HR internationalization like is repatriation, adopting organizational practices that facilitate workers retention when they return, namely through mentoring programs during the international assignment as well as career planning sessions.

These mentioned issues are very pertinent to the good functioning of the organization' Talents Management system and its continued success.

5 Conclusions

Nowadays, global organization "travel" in a new work world that requires dramatic changes in leadership, talent and HRM strategies. In this new work world, barriers between the job and personal life are completely blurred—workers are in permanent contact with organizations, product of mobile technology. Networking tools, such as Linkedin, Facebook, and others, allow individuals to a constant market monitoring, providing them the knowledge of new job opportunities in real time. Many people work in global teams, and well-succeed organizations make efforts to integrate its HR in talent attracting and development programs.

To HRM, this new world where we live requires an innovator thinking that constantly challenges the existing practices. Indeed, the present research allows us to conclude that global talent management is not Utopian in the Portuguese reality, but real, being the studied organization a good example of that.

The research main aims previously defined were clearly obtained as follows:

- Checking the existence of a formal global talent management in a Portuguese multinational.

  The studied multinational presents a Talent Management System well organized that allows them to:

  (1) Attract talent collaborators;
  (2) To take advantage of its HR strengths;
  (3) Disseminate its organizational culture;
  (4) Create an employer brand with a strong national visibility and increasing international visibility; and
  (5) Do from its talent philosophy an important source of competitive advantage.

- Empirical checking of an existent conceptual tool relative to the global talent manager enabler roles in multinational organizations and analysis of its organizational context impact.

  The existence of a corporate, talent management, and leadership development area led by a Head of Talent, reinforce the different roles that HR managers can have as global talent management enablers, and guarantee the workers development to the organization, allowing to conciliate more massive and inclusive approaches (Talent Matrix), with more customized approaches (Management and Leadership Academy) for the members of the Administrative Board and High Potentials, whose strategy is drawn for growth. Talent management processes are designed to multigeographies, respecting the need of some local adjustments, such as the adoption of Spanish language to business with Spanish people, replacing the traditional English.
- Checking the existence of the principles to the global talent management effectiveness

  The presence of the six principles defended by Stahl et al. [16] to the global talent management effectiveness—alignment with the strategy, internal consistency, cultural roots, management commitment, balance between global and local needs, and employer branding through differentiation—contributes to the organization global talent management System success.

Concluding, it is possible to consider the strategy and the global talent management system adopted by the studied organization as a success model to be followed by other organizations, always dully supported by changes imposed by each organizational context. The used model, per se, should not be copied. On the contrary, it could be used as a theoretical basis to the development of new models.

6 Limitations and Suggestions for Further Researches

As in any other study that presents a theoretical tool developed in a particular reality, the main limitations are based, essentially, in its incapacity to generalize. Therefore, and although we are in front of a well succeed global talent management System, its implementation in the diverse types of organizations and geographies could be limited, product of the exercised influence by each particular organizational context.

In this regard, in the scope of a global talent management and in a scenario characterized by a scarce national academic production, its analysis and deepening, in the most diverse organizational contexts, will be of great contribution to the deletion of possible existent limitations as well as to the development of a field increasingly actual and pertinent to HRM and the business world.

So, the following themes can be suggested to future research, namely:

(1) **Development of theoretical models of Talent Management**: Theoretical tools allow that other researchers develop or improve the existent ones at the same time they allow to better analyze all global talent management process.
(2) **Global Talent Management in SME**: SME represents the majority of the Portuguese business market. It is important that these organizations recognize the need of a global talent management and take conscious of its advantages. SME requires specific approaches of global talent management which are interrelated with the organizations own characteristics, such as the small number of collaborators, or sometimes, with a management familiar character, or with inexperience at the internationalization level.
(3) **Women and Global Talent Management**, developing issues such as glass ceiling and equal opportunities. Women still have lesser representation in executive commissions and in leadership positions, reason why it is urgent to analyze in what way women and their organizations could break the glass ceiling, comparing leadership, of thought and emotional styles.
(4) **Global Talent Management and Ethics**: Analysis of the talent programs ethic. HRM should consider the ethic involved in the global talent management systems. The workers exclusion from talent programs can be seen as a way to suppress opportunities to express the individual identity and authenticity. Management male nature also could form a threat to women progress and to equal opportunities. Global talent management should be followed by a sense of justice.
(5) **Global Talent Management and Innovation**: In a global world, innovation is seen as an imperative to organizations. In order to be competitive, organizations should take advantage from its capacity to innovate. The higher its capacity to innovate, more efficient will be the talent management strategies. However, we should take in mind that innovation requires creativity that often goes by simplicity, which, curiously, not always is of difficult

operationalization. Innovation in global talent management assumes the existence of distinguishing skills, different ways of leadership development, and talent culture.

References

1. Hubbard, Elbert. (1901). *A message to Garcia and thirteen other things*. New York: Roycrofters.
2. Mullaney, E. (2012). Talent mobility. *Magazine of World at Work, 2,* 33–36.
3. SHRM. (2011). *Workplace forecast: The top workplace trends according to HR professionals*. http://www.shrm.org/research/futureworkplacetrends/documents/11-0014wpf_posting_6.pdf. Accessed in December 18, 2013.
4. Nijveld, J. M. (2014). *Role of global talent management in organizational performance*. Third IBA Thesis Conference, July 3, 2014, Enschede, The Netherlands.
5. Tarique, L., & Schuler, R. (2010). Global talent management: Literature review, integrative framework, and suggestions for further research. *Journal of World Business, 45*(2), 122–133.
6. Manpower. (2011). *Manpower Group talent shortage survey 2011*. http://us.manpower.com/us/en/multimedia/2011-Talent-Shortage-Survey.pdf. Accessed in December 15, 2013.
7. Heidrick & Struggles. (2012). *Strategic talent management: The emergence of a new discipline*. http://www.heidrick.com/~/media/Publications%20and%20reports/HS_StrategicTalentManagement.pdf. Accessed in January 12, 2015.
8. Guthridge, M., McPherson, J., & Wolf, W. (2008). Upgrading talent. *The McKinsey Quarterly,* (1), 61.
9. Scullion, H. & Collings, D. (2011). *Global talent management*. New York: Routledge.
10. Schuler, R. S., Jackson, S. E., & Tarique, I. (2011). Global talent management and global talent challenges: Strategic opportunities for IHRM. *Journal of World Business, 46*(4), 506.
11. Kolo, P., Strack, R., Cavat, P., Torres, R., & Bhalla, V. (2013). *Corporate universities: An engine for human capital*. Boston Consulting Group. https://www.bcgperspectives.com/content/articles/human_resources_leadership_talent_corporate_universities_engine_human_capital/. Accessed in March 12, 2015.
12. Strack, R., Baier, J., & Fahlander, A. (2008). Managing demographic risk. *Harvard Business Review, 86,* 119–128.
13. Goldstone, J. A. (2010). The new population bomb. *Foreign Affairs, 89,* 31–43.
14. Rich, M. (2010). Factory jobs return, but employers find skill shortage. *The New York Times*, July 1. http://www.nytimes.com/2010/07/02/business/economy/02manufacturing.html?_r=1. Accessed in January 12, 2015.
15. Tymon, W. G., Strumpf, S. A., & Doh, J. P. (2010). Exploring talent management in India: The neglected role of intrinsic rewards. *Journal of World Business, 45*(2), 109–121.
16. Stahl, G., Björkman, I., Farndale, E., Morris, S., Paauwe, J., & Stiles, P. (2012). Six principles of effective global talent management. *MIT Sloan Management Review, 53*(2), 25–32.
17. Sparrow, P., Scullion, H., & Farndale, E. (2012). *Global talent management in professional and financial service firms: A key strategic HR challenge*. Paper presented to the IFSAM, Track 11 Strategic Talent Management and Human Capital, Limerick, Ireland.
18. Farndale, E., Scullion, H., & Sparrow, P. (2010). The role of the corporate HR function in global talent management. *Journal of World Business, 45*(2), 161–168.
19. Somaya, D., & Williamson, I. (2011). *Embracing turnover: Moving beyond the "war for talent". Global talent management*. New York: Routledge.
20. Guest, D. (1990). Human resource management and the American dream. *Journal of Management Studies, 27*(4), 377–397.

21. Teagarden, M., Meyer, J., & Jones, D. (2008). Knowledge sharing among high-tech MNCs in China and India: Invisible barriers, best practices and next steps. *Organizational Dynamics, 37*(2), 190–202.
22. Li, S., & Scullion, H. (2010). Developing the local competence of expatriate managers for emerging markets: A knowledge based approach. *Journal of World Business, 45*(2), 190–196.
23. Sustainability Report. (2014). Y company, Portugal.
24. Management Report. (2014). Y company, Portugal.
25. Performance management system manual. (2014). Y company, Portugal.

Index

C. Machado (ed.), *Competencies and (Global) Talent Management*,
Management and Industrial Engineering, DOI 10.1007/978-3-319-53400-8

MIX
Papier aus verantwortungsvollen Quellen
Paper from responsible sources
FSC® C105338

If you have any concerns about our products,
you can contact us on
ProductSafety@springernature.com

In case Publisher is established outside the EU,
the EU authorized representative is:
Springer Nature Customer Service Center GmbH
Europaplatz 3, 69115 Heidelberg, Germany

Printed by Libri Plureos GmbH
in Hamburg, Germany